Digital Consciousness

A Transformative Vision

Digital Consciousness

A Transformative Vision

Jim Elvidge

BOOKS

Winchester, UK
Washington, USA

First published by iff Books, 2018
iff Books is an imprint of John Hunt Publishing Ltd., No. 3 East Street, Alresford,
Hampshire SO24 9EE, UK
office1@jhpbooks.net
www.johnhuntpublishing.com
www.iff-books.com

For distributor details and how to order please visit the 'Ordering' section on our website.

ISBN: 978 1 78535 760 2
978 1 78535 761 9 (ebook)
Library of Congress Control Number: 2017944841

A CIP catalogue record for this book is available from the British Library.

Design: Stuart Davies

Printed and bound by CPI Group (UK) Ltd, Croydon, CR0 4YY, UK

We operate a distinctive and ethical publishing philosophy in
all areas of our business, from our global network of authors to
production and worldwide distribution.

Contents

To Tashi

Acknowledgments

First, I must repeat my acknowledgment from my first book of the many sources of inspiration for this book. Open-minded "out-of-the-box-thinking" scientists such as Eugene Wigner, John Wheeler, Gerard 't Hooft, Brian Josephson, Anton Zeilinger, David Bohm, Kip Thorne, Paul Davies, Ervin Laszlo, Fred Hoyle, Andrei Linde and Alan Guth have always challenged my worldview.

Scientists, mathematicians and researchers like Thomas Campbell, Dean Radin, Jessica Utts, Rupert Sheldrake, Rick Strassman, David Deutsch, Ed Fredkin, Konrad Zuse, Brian Whitworth, John von Neumann, Michael Cremo, Daryl Bem, Ian Stevenson, Ervin Laszlo, Stephen Meyer, Robert Jahn and the rest of the Princeton PEAR lab took great risks with their reputations by opening investigations beyond the status quo. Their efforts were both inspirational as well as contributory to my thesis.

Writers like Sidney Kirkpatrick, Michael Talbot, Graham Hancock, Arthur James, Anthony Peake, Robert Monroe, William Buhlman, Steven Kaufman and Jim Marrs were never afraid to tackle tough topics, and brought alternative viewpoints and theories to my attention.

Deep thinkers, futurists and philosophers like Nick Bostrom, Deepak Chopra, Brian Weiss, Frank Tipler and Neal Stephenson have all shaped my view of reality through their works.

Broadcasters like Art Bell and George Noory provided an outlet for many of these people to espouse their views.

All of the aforementioned people can be considered inspirational sources for my thesis.

I wish to acknowledge my *Universe—Solved!* peeps—people who have written to me over the years, exchanged ideas, joined my Forum, and been kind and steadfast supporters of my work.

You know who you are. ☺

Finally, to Tashi for her support and love, wise Eastern knowledge, long midnight discussions, help in editing and reviewing, and patience with me spending "yet another night on the book."

Introduction

There are many ways to tell a story. Broadly speaking, there are three: top-down, bottom-up and everything else. Most narratives have a message, a central theme or a point to get across. A scientific or philosophical work may have a hypothesis or theory to present and the objective of the work is to provide the evidence to support the theory.

In the top-down method, the overall message is presented first, followed by a set of rationales for that message, which we might call "the next layer down." Should any of those rationales require further breakdown for evidence or clarity, another layer of rationales is presented, and so on, methodically, until the author reaches whatever bottom level of explanation is comfortable. In the New Age world, that bottom layer is never very far down, and the authors use vague ambiguous terms like "vibrational level" and "energy field." In a rigorous scientific thesis, the bottom level may be full of equations and field-specific jargon that is difficult for the average reader to comprehend. The downside of such a storytelling approach is that the interesting points are presented up front, when the reader is still skeptical because the foundational material has yet to be presented. As the reader progresses through the work, they have an increasingly tedious task of slogging through deeper and deeper levels of detail.

The bottom-up method is certainly no better, because the most detailed material is at the beginning and may be sure to turn off the reader before any attention-grabbing context is given. *The Universe — Solved!* was written in such a manner and I had to keep attempting to tantalize the reader with a promise that "it would all come together in Chapter 7."

So, I am doing my best in this book to present my argument and foundational elements in a non-linear Tarantino-esque narrative that blends context and detail into every section of the book. It is an experiment.

I am also freely using elements of some of my blogs, because they happen to provide explanations for some of the concepts within. But rest assured, dear reader, that there is plenty of new material here to sink your teeth into!

Chapter 1

Why Is This So Cool?

It Answers All of the Most Interesting Questions:

- What is life all about?
- Why does the mere act of observation appear to alter reality?
- Is there life after death? If so, what's it like?
- How is it that objective reality doesn't exist, according to recent experiments in physics?
- Why does the universe appear finely tuned for the existence of matter, let alone life?
- What is matter? What is dark matter?
- Do parallel universes really exist?
- If a tree falls in the forest and there's nobody around to hear it, does it make a sound?
- What explains quantum entanglement? The quantum Zeno effect? The delayed choice quantum eraser (apparent retrocausality)?
- Are paranormal experiences real? How do they work? What are UFOs?
- Is it nature or nurture? Or neither? What explains drastic differences between the values and personality traits of identical twins who have the same genetics and were brought up in the same environment?

What is cool about Digital Consciousness Philosophy is that it provides a theory, framework or an answer to all of the preceding burning questions. Details are coming.

It Can Change the World

Are you tired of the impression of the world and society that is imprinted upon us by the evening news? While the preponderance of negative stories are presented in a disproportionate amount relative to the myriad uplifting narratives that simultaneously exist, defenders of the media may argue that misery sells. The press is simply giving us what we want to see. Yet, the bigger truth is that the real crimes of society are underreported—stealth imperialism, war profiteering, greed, government corruption, lack of willingness to care for the truly unfortunate, cruelty to animals, big corporate interests stomping on indigenous cultures and so on.

Interestingly, were we to all have an understanding of and belief in Digital Consciousness Philosophy, things might be quite different. In fact, at the 10,000-foot level, it isn't hard to recognize that these differences represent an evolution of humanity.

For example, external consciousness implies an existence beyond corporeal death. Evidence in the form of collected personal experiences by research scientists such as Tom Campbell and Eben Alexander indicate that this existence is lasting—effectively immortal. Imagine how that would change decisions and priorities made here on Earth. All of the money and efforts toward life extension may be redirected toward life itself. The medical industry might recognize that extending our life expectancy is not an evolutionary directive. Knowing of an immortal consciousness, we may instead focus on curing diseases and improving the overall quality of life, rather than viewing people's health as a profitable maintenance plan that extends life beyond a comfortable limit.

Purpose

Digital Consciousness Philosophy imbues life with a great deal

4

more meaning and purpose than does scientific materialism. This can give individuals a new perspective on the meaning of their personal lives. Instead of focusing their efforts on winning a hedonistic survival game based on fear and the scarcity of resources, the recognition that our purpose is to learn and evolve our consciousness can lead to significantly greater generosity to fellow humans, and true respect for other species on our planet. When the theoretical game Prisoner's Dilemma is played in an infinitely iterated mode, cooperative techniques optimize the outcome. Applying this lesson to an iterative life process, we would expect to see behavioral differences that result in an overall improvement in the quality of humanity as a whole versus a focus on personal self-interest.

Priorities

The increasing human population rapidly encroaches upon and destroys habitats for countless species of other conscious life forms, as well as using them for cruel experimental medical research. Recognition that animal consciousness is rooted in the same system that begets human consciousness would most certainly serve to eliminate the cruelty and exploitation. The materialist view implies that we are in constant competition for resources, thereby driving conflicts that cause war. But given the knowledge that we are all interconnected, would wars between groups of people based on differences in dogmatic religious beliefs, arbitrary geographical boundaries or political systems make sense anymore?

The power of intent

The digital nature of consciousness implies a probabilistic system (as evidenced clearly by quantum mechanics), which generates outcomes that can be influenced by intent. The belief that "skillful intention" can change your life, your society and your world could allow people to get out of their belief traps, and actually make a difference.

As the evidence that we exist in a consciousness-driven digital reality continues to mount, so will humanity's collective belief in this idea. Not only can the understanding of this model of our world lead to novel, unifying understandings in science but, more importantly, it can also lead to a more peaceful, harmonious, just and balanced worldview.

A Brief Overview

The detailed description of the digital-consciousness model and the way it works will be presented later in the book. But, to save the reader the angst of having to wade through all of the foundational stuff before the tasty tidbits, an overview is provided here.

Let's start with the idea of "all that there is." We typically think of this in terms of our physical reality, where "all that there is" is everything that exists in the physical universe. However, that has become a very antiquated notion over recent years, as we are now forced to consider things which do not appear to be in our reality, but for which there is ample scientific evidence. Examples include dark matter, dark energy and a huge quantity (some say infinite) of physical matter beyond the Hubble volume (the Hubble volume is that which is accessible to our observation, beyond which observation is theoretically impossible owing to the hard limit of the speed of light).

We should also consider things that may have less scientific evidence, but do have a good deal of rational philosophical and scientific thought underlying them, such as parallel realities and the so-called multiverse. And finally, there is that which is even beyond the theoretical physical, but for which there is ample anecdotal and scientific evidence, such as non-physical realms, the afterlife, the "in between" lives and the "astral plane." I hope to demonstrate convincingly that this latter category is not to be ignored, but is as real as the book in your hands. So, let "all that there is" be the sum total of... well, all that there is. In our model, let's use a big gray cloud to represent this.

But this is not a collection of the physical stuff that we might think it is. Instead, it is both pure data and pure consciousness. Although it is an expansive system, it is not infinite, but of a size that is far beyond our comprehension. Physicist Tom Campbell calls this "Absolute Unbounded Manifold" (AUM). Others have referred to it as the "Global Consciousness System."

If you are right-brained, think of it as a blank canvas, on which we can create anything—minds, experiences, thoughts, cars. If you are left-brained, think of it as a programmatic substrate, upon which we can "program" anything—minds, experiences, thoughts, cars. It is data in the sense that, at its deepest level, it is organized as bits, as binary elements. The physical nature of it is not important, both because this is simply an introduction to the concept, and because it is theoretically impossible to know its true nature. This is because it is far more fundamental than we can ever have a hope of exploring experimentally. Indeed, the idea of exploring its nature with the coarse tools that exist in our (much higher level) virtual world doesn't even make any sense. It would be like probing the atom with an imaginary sledgehammer.

I am going to refer to this as "All That There Is" (ATTI) going forward. It is tantamount to being "God," in a certain sense of the word. This "God" will be discussed later in the book.

You, the reader, have a consciousness. Your consciousness is a very small component of the consciousness of the whole System, but it is bounded. I represent this individuated consciousness (IC) by a little sub-cloud within the larger one, ATTI (obviously, not to scale). See Figure 1.1.

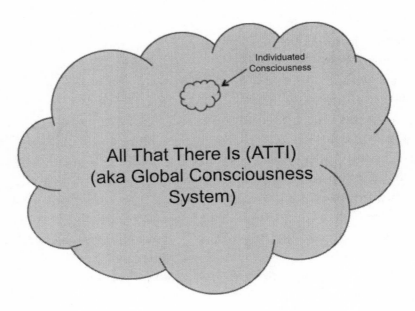

Figure 1.1

We all have ICs within ATTI—humans, dogs, ticks, fleas. The size of the sub-cloud would be relative to the complexity of the consciousness.

Another part of the system is what I refer to as the "Reality Learning Lab" (RLL). (Note: some have described life as a "learning lab" or "school" and Tom Campbell has described this reality as a "virtual reality learning lab.") Think of the RLL as a piece of virtual reality software running on the ATTI "system." The RLL contains everything that we think of as physical—galaxies, planets, cars, humans, brains, cells, atoms and subatomic particles. None of it is truly physical; rather, it is virtual data representations of those things. That's right, consciousness is separate from the brain. It does not "emerge" from complex brain functions as material reductionists would have you believe. As we shall see, the evidence supporting this is overwhelming.

Instead, consciousness is fundamental—it is the stuff of

which everything is made. Putting this all together, Figure 1.2 demonstrates the nature of reality in Digital Consciousness theory (DCT). Jim and Brandon are two individuals who exist as ICs in the global consciousness system, aka ATTI. They each have an information connection to the RLL subsystem where, along with the other ICs of 7 billion people on the planet, 500 million dogs, 30 trillion gnats, etc., they interact with each other, as well as the other artifacts in the RLL (cars, rocks, graduation hats).

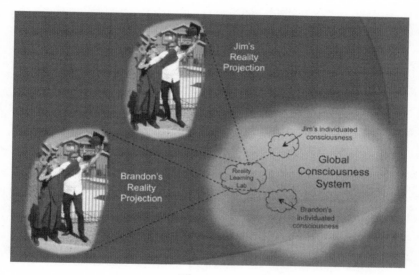

Figure 1.2

It is perfectly analogous to the experience of playing a multiplayer video game. The conscious entity is the player, but the projection of that player's reality in the game is his or her avatar plus the rest of the artifacts in the game, just as is shown in the figure.

Is it really that hard to believe? Another analogy would be a dream. Have you ever awoken in the middle of the night from a nightmare with your heart pounding? Why would your body react that way if your mind didn't believe that what it was experiencing was real?

Essentially, both the video game and the dream are projections. So, why do we think that dreams aren't real but the waking state is? There are two reasons:

1. Temporary State of Being: When we wake from the dream state, we recall the pre-dream state of the waking reality that we had yesterday. Hence, we feel that the dream is transient, a temporary excursion of our mind amidst another more permanent state of consciousness. However, as we shall see, we ultimately realize that our "normal waking state," "physical reality," "life here on Earth," whatever we want to call it, is also just a temporary state sandwiched in between a more permanent state of consciousness. Therefore, the mere fact of not having this realization or understanding is not a valid reason to believe that physical reality is concrete.

2. Consensus of Experience: When we dream, our experiences appear to be our own. There is generally no consensus established about those experiences with other conscious individuals. However, in the waking state, we have significant consensus of experience. When we are standing on the street corner with friends and a blue car drives by, we can all agree that we saw a blue car, identify the brand, the number of passengers, the speeds of the vehicle, etc. Hence, the experience feels more real and concrete, because of the experience consensus that we have with others. However, there are two flaws in this argument. First of all, the dream state is not always and completely consensus-free, as mutual lucid dreaming studies strongly demonstrate. More fundamentally, however, experiments in many fields of science show that this consensus is not 100 percent. The results of many psychological experiments indicate that we all perceive the world slightly differently. Hypnosis can cause people

to not perceive things at all, while the rest of us accept them as "being there." And finally, the field of quantum mechanics has now demonstrated conclusively that objective reality does not exist. The evidence for these claims shall be presented later.

It is easy to see how this model has tremendous explanatory power. Those experimental results that imply a lack of objective reality or a lack of conscious consensus now have a foundational explanatory construct. The power of belief and the placebo effect suddenly make sense, because the act of believing and intending is separate from the apparent physical stuff with which we interact. In fact, since that "stuff," like our bodies, is not concrete but malleable, it is easy to understand why our beliefs can mold it.

Even so, despite the incredible explanatory power of Digital Consciousness Philosophy, it is unlikely to gain immediate acceptance. This is because it typically takes 30 years or so for radical ideas to evolve from theory to acceptance. The next chapter will explain why and identify the journey that this concept will take. We can then all sit back and enjoy the ride.

Chapter 2

The Evolution of Radical Ideas

"All great truths began as blasphemies."
- *George Bernard Shaw*

"Theories have four stages of acceptance:
1. This is worthless nonsense.
2. This is an interesting, but perverse, point of view.
3. This is true but quite unimportant.
4. I always said so."
- *J.B.S. Haldane, geneticist*

George Bernard Shaw and J.B.S. Haldane were well aware of a natural human tendency to resist change. Social psychologist Robert Zajonc is noted for his development of the "mere-exposure effect," which asserts that people have a tendency to prefer things with which they are familiar. The corollary to this effect is the tendency of all organisms to exhibit a fear/avoidance response in the presence of a novel stimulus.

It isn't surprising that we might fear change or react negatively to new elements in our environment, including new ideas. After all, human evolution passed through many periods where playing it safe was the best way to pass on your genes. The risk-takers—the one who challenged the saber-toothed cat, the one who ventured out of the cave at night, the one who ate the bright red berries—would get weeded out of the evolutionary tree. As a result, we have evolved to be skeptical and to resist change. It's in our DNA.

The Expert Effect

Unfortunately, this creates a behavior pattern that discourages the acceptance of new ideas. This is especially true in the academic community, owing to the "expert effect."

"The expert knows more and more about less and less until he knows everything about nothing."
- *Mahatma Gandhi*

"preposterous"
"rocks don't fall from the sky"
- *The French Academy of Science regarding reports of meteorites in the 1800s*

"No one will need more than 637KB of memory for a personal computer. 640KB ought to be enough for anybody."
- *Bill Gates*

"There is no reason for any individual to have a computer in his home."
- *Ken Olsen, founder of the now defunct Digital Equipment Corporation*

The Scientific Method

The "Scientific Method" is a structured methodology for developing hypotheses and theories. The elements of the currently accepted method are:

1. Define a question
2. Do background research
3. Construct an explanatory hypothesis
4. Test the hypothesis by doing experiments that produce

empirical data
5. Analyze the data and draw conclusions
6. Publish the results
7. Retest, ideally by impartial peers (aka "peer review process")

I say "currently accepted" because, like everything, the scientific method is in a constant state of change or flux. It is incorrect to think that the scientific method, as currently understood, is a final "best practice."

Continuous Improvement

Continuous Improvement is a business concept whereby an organization undergoes a process of continuously inspecting their processes, products and structures, and making adaptive incremental changes to improve those processes, products and structures. The idea can easily be applied to other types of entities, such as schools, teams, governments (wouldn't that be nice?), families and individuals. As such, it is erroneous to assume that the existing processes are the best they will ever be. Has there ever been a product that couldn't benefit from some improvement? A person? A team? A country? In fact, in my humble opinion, the phrase "best practice" should simply be removed from the language. It implies that there is no further room for improvement and may influence people to become blasé about that process.

And so it is with the scientific method. It is a living process. What was thought of in the Middle Ages as ideal scientific philosophy has certainly been improved upon since that time. It would be incredibly arrogant to think that in 2016 we have it all figured out and no longer need to question the validity of existing practices and methods.

Applying an attitude of continuous improvement to

one's self can only be an outstanding practice. The scientific community would benefit from a similar philosophy, as we shall see.

Scientific methodology has certainly changed over the years. Aristotle's "Posterior Analytics" may have been the first writing to formalize a logical scientific method. Islamic scientists added experimentation to the formalism of science, while medieval philosopher Roger Bacon added independent verification, and great thinkers from the European Renaissance added concepts like a greater emphasis on causation (Francis Bacon) and logical induction. Over the years, various "demarcation criteria" have been proposed to determine what constitutes science. One such criterion, the need to establish a "mechanism," has long since been eliminated from the set, as Newtonian Gravity did not meet it. Predictability and falsifiability were new criteria promoted by Karl Popper in the mid 20th century. Yet these, plus testability, observability and repeatability, are not universally agreed upon. Neither do they all apply to theories that are commonly accepted as scientific, such as field theories and string theory. In fact, according to science philosopher Martin Eger, "Demarcation arguments have collapsed. Philosophers of science don't hold them anymore. They may still enjoy acceptance in the popular world, but that's a different world."[1]

Another important idea about science that often eludes even the most reputable of scientists (and usually the "experts") is the truism that there is never certainty in science. The probability of an idea being "true" falls somewhere between a zero (absolutely false) and a one (absolutely true).

A hypothesis differs from a theory in that it is just an early suggestion, and has not been subjected to testing and analysis of the evidence. Whereas a theory has undergone significant testing and, presumably (unless it is a bad theory), is supported by a substantial level of evidence. Note that while this rigor pushes the probability of a theory being accurate toward unity, it can never get there. To have 100 percent certainty makes something a fact or a truth. The argument that gets the theory to 100 percent would be a proof. However, proofs are the domain of mathematics, not science. Facts belong in the courtroom, not in scientific research. And truth is only in the language of philosophy. Science is about hypotheses, evidence and theories. The more substantial the evidence that supports the theory, the better the theory it is.

Despite the controversy around scientific philosophy, it should be sufficient to recognize that:

1. The Scientific Method is not a universal concept, but rather a living idea that evolves according to a process of continuous improvement.
2. It is a well-structured methodology that has millennia of debate and fine-tuning behind it. As of 2017, it's the best we have.
3. There is no such thing as proof, truth or fact in science, just evidence.

Grounded with this objective overview of science, we can now turn to the eccentricities of the cult of experts, as they apply to an objective treatment of the scientific method. It is my contention that:

1. Hypotheses that upset existing theories are less likely to be accepted than ones that simply build upon pre-established foundations.
2. The more radical the hypothesis, the more vehemently it is attacked and the longer it takes for acceptance.
3. Hypotheses that upset the scientific apple cart typically take 30 years, give or take a decade, to achieve mainstream acceptance.

One wonders if the reason for this is that it takes 30 years for an expert, threatened by a new idea, to get to the point in his or her career, where they simply no longer care, while newer scientists have been able to integrate the new idea into their worldview without threatening their livelihood.

The "30 Years to Acceptance" Syndrome

To cite one specific example, let's look at the idea of "cold fusion." Nuclear fusion is a nuclear reaction whereby lightweight atomic nuclei fuse together to form heavier nuclei, generating excess energy in the process. It is generally believed that stars create their energy in this manner, by having a continuous nuclear fusion reaction at their core. Until recently, it was thought that the only possible way to create a nuclear fusion reaction was to heat the fuel components to extremely high temperatures, such as millions of degrees Celsius. In 1989, world-renowned electrochemists Dr. Martin Fleischmann and Dr. Stanley Pons reported anomalous heat production accompanied by common nuclear reaction byproducts such as neutrons and tritium (an isotope of hydrogen) in a desktop experiment at room temperature. Dubbed "cold fusion" at the time, their announcement was met with extreme skepticism. Douglas Morrison, a physicist from the European Organization for Nuclear Research (in French, Conseil Européen pour la Recherche Nucléaire, or CERN), referred to their work as "pathological science" and said "The results are impossible."[2] Steven Koonin, then professor of physics at the California Institute of Technology (Caltech) (and later the Under Secretary of Energy for Science at the US Department of Energy), commented "we're suffering from the incompetence and perhaps delusion of Drs. Pons and Fleischmann."[3] Scathing magazine articles were written and the scientists were ostracized for decades by their closed-minded peers. Some of the harshest criticism came from the Massachusetts Institute of Technology

(MIT), who may have felt the threat of losing millions of dollars of federal "hot fusion" funding if a simple tabletop experiment could demonstrate net energy creation that their hot fusion program could not.

Within a short period of time, various research institutes around the world attempted to reproduce Fleischmann and Pons' results. Some experiments showed no hallmark signs of a fusion process. Others demonstrated some excess heat and still others resulted in anomalous fusion byproducts. But the damage was done. Cold fusion had become a joke; in reality, it was adopted as a metaphor for bad science or pseudoscience. Many researchers refused to touch such a tainted area of research for fear of damaging their reputation. So, cold fusion research went underground. In fact, a very clever tried-and-true technique was employed—change the name to save face. Cold fusion became Low Energy Nuclear Reactions (LENR).

Between 1989 and 2004, there were over 15,000 replicating experiments done in the newly dubbed field of LENR at such prestigious institutions as MIT, NASA, the United States Department of Energy, the University of Chicago, Osaka University and Toyota. According to science researchers Steven Krivit and Nadine Winocur, the reproducibility rate became as high as 83 percent.[4] In 2012, during a colloquium on LENR at CERN (one of cold fusion's original and harshest critics), it was declared "The effect described by Fleischmann & Pons in 1989 is confirmed." Further, "The quality of experiments worldwide performed is so high and the results obtained so widespread/reproduced, that I believe it is the time to start an International Research Program to boost the results."[5]

So, in the cold fusion case, it took about 23 years to work through the first two stages of J.B.S. Haldane's Four Stages of Acceptance (see the beginning of this chapter). Cold fusion is still by no means accepted in the scientific mainstream.

Neither is this trend improving as we become more

"enlightened." In 1827, Georg Ohm published his now famous theory of electrical resistance (now known as Ohm's Law). It was met with harsh criticism, even to the point where the German Minister of Education said that "a professor who preached such heresies was unworthy to teach science."[6] As a result, Ohm lost his job and landed on hard times. It wasn't until 1852 that he was appointed to a university teaching position—25 years, no different from today.

In 1879, amateur archaeologist Marcelino Sanz de Sautuola discovered and then published a statement that cave paintings in the Altamira cave in Spain appeared to date to the Stone Age. The French archaeology establishment ridiculed his findings and accused him of forgery. Sautuola died in disgrace in 1888, but was vindicated in 1902, when the scientific community retracted their opposition—23 years.

In 1933, Swiss astronomer Fritz Zwicky proposes a concept of "dark matter" to explain anomalies observed in the motions of galaxies. His idea was ignored by the scientific community for decades. German astronomer Walter Baade referred to him as "mad," others as an "irritating buffoon." Zwicky died in 1974, but had been vindicated in 1973, when Princeton astronomers realized they needed dark matter to complete their model of the universe—40 years.

So as can be seen, there doesn't seem to be a trend toward the shortening of the adoption cycle of radical new ideas. Instead, we consistently seem to require 20 to 40 years for novel theories to reach the mainstream, even when they have solid evidence behind them.

The Technology Adoption Lifecycle

In the high-tech world, a related concept called the technology adoption lifecycle has its roots in the same set of fears. Based on the book *Diffusion of Innovations*, sociology professor Everett

Rogers identified the pattern of adoption of new innovations or ideas (see Figure 2.1). He noticed that there are typically very few individuals, about 2.5 percent of the population, who are willing to adopt and experience the new ideas or innovations. The next group, called early adopters, comprises about 13.5 percent of the population. Then comes the majority, followed at the end by the 16 percent of us who are the laggards.

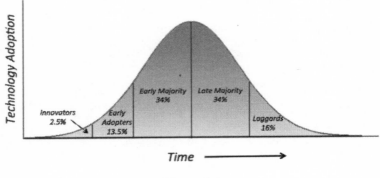

Figure 2.1

The technology cycle is usually faster than the scientific cycle, typically occurring over the period of 5 years or so, probably owing to the fact that the adopters are rushed by forces of ever-shortening technology support cycles as well as the "keeping up with the Joneses" effect. In addition, technology adopters don't have that much to lose except the cost of the new product and the time it takes to learn to use it, whereas there is much more at stake in adopting a new theory of science or reality; namely, your entire worldview.

A New Radical Idea

The idea presented in this book (as well as other books and papers written on the subject) is radical. It falls squarely into the category of "ideas that scare people" and hence, will require at least 30 years for the idea to achieve mainstream acceptance. However, it is more than a theory of science. To be sure, it makes use of the scientific method for establishing evidence, causal relationships and an analytical approach to its conclusions. It has explanatory and predictive power in various fields of science, such as quantum mechanics and cosmology. However, it also belongs in the domain of philosophy.

Interestingly, philosophy was much more tied to science in the past (hence, the application of the degree "Doctor in Philosophy," aka PhD, to the sciences). And new philosophical ideas don't follow the common time frames of technology or science. Instead, they may need more than 30 years to gain acceptance. There is no rush to philosophy; it is just there to make sense of things, to make your life better. But it is not a religion. There is no dogma, no rules and no hierarchy of leadership. These are concepts designed to keep people in control, not to enlighten them.

Although its roots can be traced back to Plato's Cave allegory, with no small influence by people like Konrad Zuse in the 1960s and Philip K. Dick in the 1980s, I would mark the general genesis of the full formulation of the idea in the year 1999, for reasons that will be described in the next chapter. Hence, as this book is being written, it is 2015 and we are halfway through the scientific adoption cycle. Indeed, I have seen a healthy progression of acceptance in scientific circles, first occurring in fringe publications and late-night talk radio, and then tangential ideas innocuously dropped in articles published in the mainstream media. More recently, ideas around simulations and virtual digital realities have even received occasional support from well-respected and established scientists. The early adopter is aware of it and the early majority is starting to pay attention.

What is it?

Chapter 3

This Radical Idea

"There are more things in heaven and earth... than are dreamt of in your philosophy."
- *William Shakespeare, Hamlet (Act I, Scene 5)*

The idea I am talking about is Consciousness-driven Digital Reality.

It is based on four distinct tenets:

1. Consciousness is fundamental and primary. This means that consciousness is not simply an artifact of the complexity of a brain. Instead, it is fundamental to the universe (or whatever we shall call "all that there is"), and more the source of our experience of reality and even the source of reality itself.

2. All matter is information; all forces are rules about how data interacts.

3. The reality we experience is illusory, a simulation of sorts, designed for us to learn and evolve our consciousness.

4. The "system" is digital and consists of, at a minimum, the aggregate of all individuated conscious entities, plus the learning lab, and is driven by a fundamental rule of continuous improvement.

Before I go any further, I must stop and give credit where credit is due—to thousands of years of independent thinkers who laid the foundation for this new idea.

Credit is Due

- *Historical Philosophers*: Plato, Zhuangzi, Descartes, Jung and the Hindu Maya Illusion all questioned the solidity of reality, contributing thought leadership to the idea of an illusory reality. Plato's Allegory of the Cave explores the idea that we build realities based on what we experience through our senses, which may only be a small fraction of the truer reality. Similarly, Edwin Abbott Abbott's novella *Flatland: A Romance of Many Dimensions* modernizes Plato's allegory by considering what we might experience if our senses are limited to a set of dimensions that is less than true reality. The so-called Dream Argument dates back to the 4th century BCE, when Chinese philosopher Zhuangzi dreamed of being a butterfly and pondered: "Now I do not know whether I was then a man dreaming I was a butterfly, or whether I am now a butterfly dreaming I am a man." French philosopher and scientist René Descartes penned his *Meditations on First Philosophy*, which questions all that we perceive through our senses and considers what little can actually be known with certainty.

- *Physicists*: Eugene Wigner, John Wheeler, Gerard 't Hooft, Brian Josephson and Anton Zeilinger are physicists who pushed (and continue to push) the boundary of physical reality, contributing to supporting evidence for tenet No 3. Their view may be a little different than that of the historical philosophers previously mentioned, as it has both the advantage and disadvantage of being somewhat grounded with experimental scientific evidence for the lack of an objective reality.

- *True Consciousness-driven Digital Thinkers*: Tom Campbell (physicist, author of *My Big TOE*) and Steven Kaufman (author of *Unified Reality Theory*) are two individuals who fully embraced all 4 Digital Consciousness tenets, and independently developed a comprehensive theory of the foundations, functions and evolution of reality. I humbly add myself to this camp and feel that my contributions to this effort consist largely of consolidating a huge wealth of evidence for the theory, developing detailed explanations for anomalies in quantum mechanics and offering a statistically sound model to explain it all. We all developed our ideas on this model independently.

- *Experiencers*: Robert Monroe, William Buhlman and others are researchers who have had the gift of the ability to explore the greater consciousness and realms beyond the so-called physical.
- *Digital Physicists and Mathematicians*: David Deutsch, Ed Fredkin, Konrad Zuse, Brian Whitworth and John von Neumann are physicists and mathematicians who have been thought leaders in digital physics, and supporters of tenet No. 2.
- *Modern Philosophers*: Nick Bostrom (philosopher from Oxford University and author of *The Simulation Argument*), David Chalmers and others who have embraced the simulation model.
- *Pop Culture*: Philip K. Dick, the Wachowskis Brothers and others are popularizers who have actually done a great service to bring simulation scenarios, life after death and paranormal topics into public consciousness.
- *Free-thinking Scientists*: Daryl Bem, Ian Stevenson, Dean Radin, Ervin Laszlo, Stephen Meyer and many other scientists deserve a great deal of credit for having the courage to explore the paranormal and publish their findings in an often hostile scientific culture.

A Word About Words

The purpose of words is to convey ideas to others. Sometimes, we encounter topics for which existing words are insufficient. This is probably one of them. Words like God, consciousness, illusion, physical, virtual, simulation and reality are extremely difficult to use casually, because everyone has a slightly different understanding of what they mean. It honestly isn't something worth worrying about. If you don't like one word, substitute it in your mind for something that resonates better with you. Ultimately, the purpose of this book is to attempt to convey some ideas that may cause people to think differently and enrich their lives in some way.

That said, it may be useful to state some assumptions about certain words that I will be using in this book, along with some definitions that help the reader to appreciate the arguments made herein.

- *Materialism:* "A theory that physical matter is the only or fundamental reality and that all being and processes and phenomena can be explained as manifestations or results of matter" (per Merriam-Webster) *[author's note: It is refreshing to see that Merriam-Webster properly recognizes that this is just a theory].*

- *Reality:* By convention, this has traditionally been the physical reality that we normally experience during our waking state. However, I think that it is more accurate to describe "reality" as a general category of environment, which appears to have all of the elements needed to carry out a meaningful existence. So, therefore, a well-designed computer simulation would generate a type of reality, as would a comprehensive dream state, and, of course the physical waking environment here on Earth. In addition, the so-called astral plane, or "heaven," or the environment that we typically encounter after death, are realities. In fact, to some extent they are more real than our physical waking state reality.

- *Physical Reality (PR):* By convention, this is normally used to mean the physical waking state reality that we all experience when we are not sleeping, meditating, under the influence of hallucinogenic drugs, or experiencing OBEs, NDEs or mystical experiences. Isn't it ironic how many environments we have to exclude to clarify what the physical reality is? But, to be more accurate, what we call physical appears to be virtual. So, henceforth, I will use the term "waking reality" to mean what is traditionally thought of as physical reality.

- *Waking Reality:* See Physical Reality.

- *Virtual Reality (VR):* By convention, this has typically been used to indicate an environment or "reality space" that is layered on top of waking reality, with the assistance of reality-generation tools such as computer simulations and VR goggles. In truth, we have no idea what the *real* physical reality is and so everything else is virtual. Therefore, I use virtual in a relative sense. Experiencing a simulation that is layered on the physical reality makes it virtual to that physical reality. If we were in turn experiencing a simulation within that simulation, we would say that this "second level" simulation is virtual to the "first level" one.

- *Real:* "Real" in the new context can only mean "that which we actually experience." Therefore, a dream is real, as is an OBE and our waking reality. It is all real as long as we are experiencing it. What our neighbor experiences, however, cannot be real to us.

- *The Universe:* The word "Universe" is used in many different contexts and, unfortunately, recently it has been demoted in significance, not unlike poor Pluto, the celestial body formerly known as a planet. In the materialist world, the Universe is everything. But then along came ideas like parallel universes, Hilbert Space and the Multiverse, and suddenly "Universe" didn't seem so all-encompassing.

Then there are those who use Universe to mean "God, except that I don't believe in the traditional Judeo-Christian concept of God but instead in a vague new-agey idea that the Universe looks out for me," as in, "I am going to ask the Universe for advice." However, for clarity, I am only going to use Universe to mean the apparent collection of physical material that extends as far as the Big Bang was able to send it. In other words, what most people think of when they say "Universe." Tom Campbell refers to this as Physical Matter Reality, for those who wish to relate these ideas to his *My Big TOE*.[1]

- *Multiverse:* Max Tegmark has probably done the most in the mainstream science and math community to define what is meant by Multiverse. In that definition, he includes all other forms of Physical Matter Realities that may be "out there," including other "Big Bang-like" universe bubbles, a Hilbert Space of parallel worlds generated by quantum mechanical choices and mathematically failed universes that have to exist in order to solve the materialist dilemma of the finely tuned universe anomaly. It does not, however, include larger constructs, such as the place we seem to travel to when we die and the higher-level reality that might be generating all of the rest of the Multiverse stuff via a simulation-like experience.

- *ATTI:* All That There Is is the true foundation of reality. Being digital (as will be shown) and highly organized, it is pure consciousness. It is a system that contains all of us, as well as mechanisms for us to experience reality and evolve our consciousness. The universe (and if it exists, the Multiverse) are simply projections of experience and, mostly likely (again, as will be shown), a miniscule part of this system.

- *Subjective:* What we experience.

- *Objective:* What exists independent of experience. Quantum mechanics and other research have effectively proven beyond a reasonable doubt that objective reality does not exist. However, most of the rank and file of scientists have not yet come to terms with that.

In addition, there are many instances of the use of exponential notation in this book. For those who are unfamiliar with this form of expressing large numbers, it works like this:

2E12 is the same as $2*10^{12}$, which is the same as the number 2 followed by 12 zeros, or 2,000,000,000,000, also known as 2 trillion. Those are just many different ways to express the same concept.

Closely related to words are *models*. A model is simply a way of describing something. The number "2" is a model for the idea of having two things. General relativity is a theory of space, time and matter that can be understood equally well using different models. The equations of relativity can be used to derive the relative speeds and masses of moving objects, and the dilation of time and space, such as:

$$x' = gx - gbct$$
$$y' = y$$
$$z' = z$$
$$ct' = gct - gbx$$

Another model describes the exact same transformations, using matrices instead of linear equations.[2]

But a model of warped space (like what a heavy rock does to the surface of a trampoline) can be equally effective in describing what happens to speeds and masses of moving objects as they travel near a massive object.[3]

All are models that are equivalent in describing the same

28

effects.

People use a model like "holographic" to describe a couple different things that relate to the nature of reality:

1. The idea that deep down, in a way that is not apparent to normal senses or means of measurement, things are inter-connected.
2. The idea that information about all things exists at every point in space.

I only bring up these ideas to emphasize the point that models, metaphors, theories and descriptive words are just that—models, metaphors, theories and descriptive words. People use them to help explain ideas and experiences. But they are nothing real.

Consciousness-driven Digital Reality

Now let's examine the four tenets in more detail. Comprehensive evidence for each of these will be presented in Chapters 6 and 7, so feel free to skip ahead if you aren't buying it.

1. *Consciousness is fundamental and primary.* The mystery of consciousness has occupied the thoughts and writings of philosophers for millennia. And, although modern science attempts to develop explanations for every unknown, we are still no nearer to a clear understanding of consciousness than Plato was. In fact, one might argue that the longer the explanation for something eludes us, the more fundamental that thing is. Just as water is fundamental to a fish's reality, so is consciousness fundamental to our reality.

"With no light and only a dim awareness, the fish knows nothing of water. Water just is, has always been, and is taken for granted. The fish does not ponder the nature of water, it swims in it. We swim in an ocean of consciousness. We are not aware of the ocean, but only of our local interactions with it."[4]

- *Tom Campbell*

"Meditate, vibrate upon the Lord; immerse your mind in Him, like the fish in the water."

- *Guru Nanak, founder of the Sikh faith*

Consciousness doesn't "wink out" when electrical activity ceases in the brain. As we shall see in Chapter 7, the evidence is incontrovertible that consciousness can continue to exist in the complete absence of neural activity. Just as the existence of a single white crow would disprove the statement "all crows are black," the existence of a single instance of consciousness existing beyond a functioning brain disproves the statement that consciousness emanates only from brain function. Given that a 1992 Gallup Poll estimated that about 13 million Americans have experienced a near-death experience (NDE),[5] even if a small fraction of those experiences defy the logical explanatory argument of a dying brain (as a significant fraction of them indeed does), we have way more than the single white crow necessary to disprove the materialist myth that consciousness emanates from the brain.

In fact, why do we need to disprove something that has never had a single shred of evidence in the first place? For evidence, some have made several fallacious arguments, such as experiments utilizing MRI technology that show that "neurochemical processes produce subjective experiences."[6] The problem here is the word "produce." Does the neurochemical process *produce* the experience or does the neurochemical process *enable* the experience? If I were to attach an oscilloscope input to the circuitry of a television set and make the observation that a movie appears on the screen when we observe voltage signals in the television circuitry, should I conclude that those signals are the source of the movie? Of course not. All they did was take part in the facilitation of the decoding of the movie into a signal that could be observed on the screen. The source of the movie was a broadcast center thousands of kilometers away. In our analogy, electrochemical signals in the brain facilitate the experience, but are not necessarily the source of the experience. And once we look at the very real evidence that disproves neurochemical causes for the effect of experience, we can remove the word "necessarily" from the last sentence. Sadly, many scientists don't follow the scientific method of following the evidence where it leads and instead cling to the materialism theory that is routinely taught in schools as if it were a fact.

2. *All matter is information; all forces are rules about how data interacts.* There is actually no evidence that there is really anything solid at a fundamental level. I like to use the word "stuff" to denote that which is theoretically indivisible. The Greeks coined the term atom (atomos), which means exactly that. Even until the early 1900s, scientists thought that atoms were indeed the fundamental building blocks of matter. In 1909, Ernest Rutherford discovered that the atom was actually mostly empty space with a "solid" nucleus when he and his colleagues fired alpha particles at atoms, and some of them bounced directly back. He said of the experiment, "It was almost as incredible as if you fired a 15-inch shell at a piece of tissue paper and it came back and hit you."[7] Quark theory and string theory have progressively pushed the understanding of matter to more and more tenuous models. The direction this trend of discoveries is taking on the nature of matter is that there is ultimately no "stuff" at all.

The popular objection to the idea of matter as information is that we *feel* stuff when we touch things. But what are we actually experiencing? Even with the Rutherford model of the atom, we don't actually make contact with anything when we knock on a door. If we were to visualize exactly what is happening at the atomic level in extremely slow motion, what we would see is the molecules in the surface of our knuckles getting ever closer to the molecules in the surface of the door. The two sets of molecules never actually make contact. Instead, the closer they get, the greater the repulsive force of electromagnetism that will exist between them. Imagine having two very powerful magnets and attempting to push the north ends of each together. The repulsive force is easily felt. It is this repulsive force at the atomic level that makes you feel the slight pain sensation on your knuckles. But no "stuff" has to exist to make this happen.

The next argument might be that the force itself is carried by particles, which are a form of stuff, and it is the aggregate of those particles that we are feeling. Possibly, but certainly not necessarily. All that we really need in order to experience the feeling in our knuckles when we knock on the door is something (a force, a particle, a transfer of information) that tells the molecules at the tips of our knuckles that they have been repelled, and our nerve endings, electrochemical signals to the brain and computational processing takes care of the rest.

Some people react negatively to the idea of everything being composed of bits. But this is only because, to this point, the things in the world that we most often think of as binary are technological — like smart phones and computers. So we associate the idea of being digital with cold, calculating technology. But there is nothing cold about flowers, music, love and emotions, and there is no reason for them to be analog versus digital in nature.

In fact, as we shall see, the idea that matter is simply information is not only self-consistent with everything that is observed in nature, but it also solves an incredible number of anomalies that will continue to dog science until they acknowledge this new model. Much more to come in Chapter 6.

3. *The reality we experience is illusory, a simulation of sorts,*

designed for us to learn and evolve our consciousness. There are actually two elements of this tenet. One is that our reality is illusory and the other that it has a purpose. Let's tackle the former idea first. When we dream, we believe it is real. If our memories were erased and a new set of memories implanted in our minds, and we were subject to a fully immersive virtual reality experience, we would be in a simulation and not know it. In fact, it is absolutely impossible to know for sure that we are not. The idea of a simulation is not as kooky as it sounds at first. The world's leading physicists have shown that, beyond a reasonable doubt, there is no objective reality. Consciousness appears to be an integral part of the creation of not only our subjective reality, but also of the consensus reality experienced by others. If this is true, and consciousness is separate from the brain, as argued previously, then a simulation is almost assured.

The use of the word "simulation" can certainly be debated, as it is not the most apt description of the kind of experience that I am presenting. Simulation, as it is generally used, implies something artificial, such as a virtual reality simulation or a flight-training simulator. However, there is nothing unreal about our experiences.

4. *The "system" is digital and consists of the aggregate of all individuated conscious entities, plus the learning lab, and is driven by a fundamental rule of continuous improvement.* This idea puts it all together. Not only is matter digital, but so is consciousness. It is interesting that some have an adverse reaction to this idea and seem to think that it makes our decisions pointless or that it implies solipsism. Neither is necessarily true. It really shouldn't change how you carry out your life. If anything, the recognition that this apparent reality isn't all that there is might even make one act more honorably, inasmuch as there is a point to it all. Your mind is your mind, and your awareness and experiences are still real. If the point is to learn and evolve, as it seems, one should continue to make the best decisions that help and support

those who you love. Because their awareness and experiences are real, too. And inferring solipsism is simply a mistake. Multiple individuals with free will interacting in this reality, whether you call it a simulation or something else, is simply not solipsism. Much more about this in Chapter 7.

Taken as a whole, these four concepts together form a very powerful framework that completely explains all aspects of reality, including the most challenging philosophical and scientific conundrums.

Chapter 4

Philosophy, Science and Theories of Everything

"The first gulp from the glass of natural sciences will make you an atheist, but at the bottom of the glass, God is waiting for you."
- *Werner Heisenberg*

Depending upon your mentality, outlook on the world and conditioning, you may be tempted to approach the Digital Consciousness idea with the mindset of "but is this science?"

But why do we even ask such a question? What does it even mean for something to be "scientific?"

To begin to explore the answers to these questions, we need to consider what it is that we care about when considering this theory. If you are like me, we care about the likelihood that the theory represents truth and how much truth there is in it. Science does provide a framework for accumulating supporting evidence and eliminating aspects of a theory or an entire theory via conflicting evidence. To that end, the scientific method can be very useful in helping us to validate or refine our theory.

However, science can't say much about existence, consciousness, awareness, spirituality, ecstatic experiences, or even truly knowing something. Descartes said, "I think, therefore I am." Actually, being French, he said *"je pense, donc je suis"* in his *Discourse on the Method* in 1637, and later wrote the Latin *"cogito ergo sum"* in *Principles of Philosophy*. "I think, therefore I am" was a translation, as is perhaps a clearer alternative form "I am thinking, therefore I exist." To know that you exist may actually be the closest thing we have to subjective certainty. And yet, there are no scientific experiments that can be done to validate it. It is simply "knowing."

What is Science?

In actuality, the modern "scientific method" is only a few hundred years old; although many sciences, such as medicine and astronomy, have been practiced for thousands of years. No two philosophers of science will agree on what constitutes science.

A modern, and commonly referenced, scientific method consists of the following steps:

1. Ask a question
2. Do research
3. Construct a hypothesis
4. Test the hypothesis via experiments
5. Analyze data and draw conclusions
6. Report results

Depending on the conclusions drawn, a feedback loop in the process may be necessary, in order to refine the hypothesis and create new experiments to generate more data.

As a hypothesis becomes stronger, owing to the amount of supporting evidence for it, it enters the realm of a "theory."

And that's pretty much the extent of it. Note that there is no concept of "proof" in science. Proofs are in the domain of mathematics. In fact, there are no absolutes in science. Science deals only with hypotheses, evidence and theories. The more evidence that supports a theory, the stronger the theory should be considered to be.

As a methodical process, the scientific method is an excellent tool in providing a framework for theory refinement. However, it should by no means be considered a universal or fundamental concept. It's rather obvious that it isn't given the many refinements applied over the centuries, but it is sometimes hard to recognize that everything we base our society on, including ideas about science, are simply fluid and temporary ideas that fit a contemporary context. The coming realization that there is no such thing as 100 percent fixed objective reality, for example, will certainly reshape our views on what science is and how it should be conducted.

Another aspect of the definition of science is the concept of demarcation criteria. These are attributes that can be used in an attempt to categorize theories and concepts into science or non-science. Over the years, various demarcation criteria have come into and out of favor. Some have asserted that for something to be "scientific" it must meet these criteria and, as a result, science "bigots" regularly trot these criteria out to assert that ideas, which disrupt their cherished worldview, are "pseudo-science." A partial list of commonly used demarcation criteria follows:

- Testability: Is it possible to derive a test that furthers or refutes the hypothesis or theory?
- Falsifiability: Is it possible in theory to determine that the hypothesis or theory in question is false?
- Observability: Is the theory observable?
- Predictability: Does the theory make currently untested predictions that can later be tested for validation?
- Repeatability: Are the outwardly observable aspects of the theory consistently repeatable?
- Mechanism: Is there an underlying physical material cause behind the observable aspects of the theory?

The Mechanism criteria has long since been discredited owing largely to the fact that theories of gravity (aka vortex gravity) based on an underlying mechanism (ether) fell out of favor when the evidence mounted to support a field-based theory. The concept of an ether-filled universe was ultimately dealt a death blow from the Michelson-Morley experiment. This experiment, named after scientists Albert Michelson and Edward Morley, was conducted in 1887, and used an interferometer to measure tiny differences in the speed of light propagating with the (supposed) ether and the speed of light going against the ether. Since no differences were noted, it was apparent that there was no such ether. I find it interesting that the results of a single experiment can make a scientific demarcation criterion obsolete.

In fact, just as there is no universal list of criteria that defines life, there is also no universal list of criteria that defines science. String theory is generally accepted as a bona fide scientific theory and string theory curricula exist in all of the most reputable university physics programs. But string theory is neither testable nor falsifiable. Who would tell physicists Edward Witten, Leonard Susskind, Brian Greene and Michio Kaku that their field of research is pseudo-science?

What about observability? Psychology is an accepted field of applied science. Yet, the mental states on which psychology is based are not directly Observable, only the resulting behaviors. Neither are they repeatable. In evolutionary biology, the common practice of inferring past mutations despite the lack of fossil evidence is certainly not following an Observable practice. And pretty much all theories based on fields are not Observable, except in the macroscopic sense. The Big Bang Theory, especially the inferences around the early epochs and inflation, is neither Observable nor Repeatable.

And then there is the Many Worlds Interpretation (MWI) of Quantum Mechanics. David Raub conducted a poll of 72 "leading cosmologists and other quantum field theorists" in 1995 and found that 58 percent of them agreed with the statement "Yes, I think MWI is true."[1] And yet, that theory is neither testable, falsifiable, observable nor predictable.

[Note: I fully acknowledge that some may argue these assessments inasmuch as what seemed impossible yesterday (e.g. teleportation) is

routine today and what seems impossible today (testing MWI) may very well turn out to be possible tomorrow.]

Paul Feyerabend, the late professor of philosophy from the University of California, Berkeley, argues in his book *Against Method* that there should be no strict rule-based methodology to science and that such rules simply restrict scientific progress.[2]

Digital Consciousness — Science or Philosophy?

How does Digital Consciousness fare in the test of what constitutes science?

- Testable: Yes. At Fermilab, the US's self-designated "premier particle physics laboratory," an instrument called the Holometer is being developed to study and test the quantum nature of space. The other aspect of our theory, immortal consciousness, will be tested by each and every one of us when we die. In addition, rigorous analysis of corroborating evidence of past-life experiences represents valid scientific testing.

- Falsifiable: At this point, it does seem like Digital Consciousness may not be falsifiable. On the digital front, if any experiment designed to detect the discrete nature of space (such as the Holometer at Fermilab) comes up with a negative result, it could always be because the true resolution of space is much finer than the experiment can detect. On the consciousness front, it would also seem to be impossible to determine for certain that consciousness emanates strictly from brain function.

- Observable: Most definitely. As this book will show, many have experienced directly the separate nature of consciousness. And, theoretically, a sufficiently sensitive experiment could confirm the digital nature of space and matter.

- Predictability: Yes. Keeping in mind that while a theory may make predictions, the confirmation of a given prediction can never 100 percent confirm the theory (otherwise it would be a fact, not a theory, which doesn't exist, as previously argued). Digital Consciousness can make certain predictions. For example, since the system always evolves to more profitable outcomes, there could never be an apocalyptic event. This "evening effect" is explored in my first book, *The Universe—Solved!*, and can be summarized by the argument that the state of our reality will neither trend toward disaster nor utopia, despite a statistical likelihood of doing so. The implied "state machine" nature of fundamental reality, currently fully able to explain such quantum anomalies as entanglement and the quantum Zeno effect, may be exploited to predict other anomalies. Many other predictions will be covered later in the book.

- Repeatability: Yes, but we need to look at this concept a little closer. To be able to repeat an experiment and get exactly the same results—if, by "results," we mean detailed data points—I would argue that this rarely, if ever, happens in any case. All results have error bars, owing to noise, limitations of the measurement systems and other unknown aspects of the experiment, so it is often impossible to get exactly the same data points. In addition, experimenter's bias and "observer-expectancy effect" can preclude consistent results from the same experiment done by different experimenters. Thus, "same results" or "similar results" means statistically the same or similar. To bring this point to a clear example, consider a telepathy experiment. As a subtle effect, which can easily be explained by Digital Consciousness, one will never get consistently repeatable data points. However, under similar experimental circumstances (for example, same environment, subjects selected in the same manner, perhaps even at random), the statistical outcome should be similar. For example, if a particular experiment shows a .5 percent bias toward positive results with an odds against chance of 1E-7, and subsequent experiments show similar positive biases within the error bar of the experiment, that can be said to be repeatable.

So, by 80 percent of the standard scientific demarcation criteria, Digital Consciousness theory (DCT) can be considered to be scientific, certainly more so than string theory or the MWI theory of quantum mechanics.

Even disregarding the controversy over scientific demarcation, science isn't for everything. Have you ever read a scientific article explaining what love is? If so, I bet it left you cold. Does that mean that love isn't real? Of course not. Just that love is something that probably doesn't need to be analyzed using a scientific method. But that doesn't make it worth talking about, writing about, understanding and experiencing.

So, if it is important to you, the reader, that Digital Consciousness be considered scientific, you can rest assured that it may be considered so. And if you don't care about arbitrary and short-lived definitions of science, and care more about the evidence for a greater truth that underlies everything about our reality and defines what life is all about, call it philosophy.

Theories of Everything

The commonly used expression "Theory of Everything" (or TOE, for short) is actually somewhat of a misnomer. As used in scientific circles, a TOE is really more of a framework that supports all observable laws of physics, including other theories such as general relativity and quantum mechanics. Past and present proposed physics TOEs include the Grand Unified Theory (GUT) and string theory.

"The more important fundamental laws and facts of physical science have all been discovered, and these are now so firmly established that the possibility of their ever being supplanted in consequence of new discoveries is exceedingly remote... Our future discoveries must be looked for in the sixth place of decimals."
- *Nobel Prize Laureate Albert A. Michelson, 1894*

As Albert Michelson made the mistake of thinking only in the context of his time, so do today's physicists, who think that string theory or some other competitive TOE will be the final TOE in physics. Unfortunately, there are many things wrong with this way of thinking. First of all, who decided that physics is the sole domain of TOEs? Physics will never explain consciousness. But isn't the so-called hard problem of consciousness a fundamental mystery that any true "theory of everything" should be able to address? Secondly, if we have any chance of discovering a true framework to explain ATTI, we need to get out of the mindset of thinking only in terms of our own time and culture.

Digital Consciousness can certainly be described as a TOE framework and a very comprehensive one at that inasmuch as it supports not only all foundational elements of physics, but metaphysics as well. No other theory can make that claim. The rest of this chapter will develop the powerful logic behind that claim.

Deductive, Inductive, Abductive Reasoning

A standard scientific method of developing a theory is via the logic technique of abductive reasoning. This quick aside explains the different kinds of reasoning:

- Deductive: Deductive reasoning infers specific conclusions from general rules or assumptions. As an example:

- o All swans are birds (general principle)
- o Fred is a swan
- o Therefore Fred is a bird (specific inference)

In science, deduction is used to go from theory to prediction and test. So, if the theory is true, then its predictions should be affirmed through experimentation. If they are not, the theory would need to be modified or the logic behind the prediction re-examined.

- Inductive: Inductive reasoning infers a generalization from specifics. For example:
  - o 100 observations of individual swans indicate that each is white (specific observations)
  - o Therefore, it may be inferred that all swans are white (general inference, theory)

In science, inductive reasoning is used to form hypotheses and theories. Note that the general inference is not necessarily true. So, for the purposes of science, it is better not to include definitive statements. In the above example, all it would take is a single black swan to disprove the theory. However, if the theory is stated "most swans are white," then the theory has a better chance of standing (that is, until huge colonies of black swans are found that exceed in size the previously known population of white swans).

- Abductive: Abductive reasoning starts with an incomplete set of observations and proceeds to the likeliest explanation. For example:
  - o Fred is a black bird
  - o Fred is shaped like a swan
  - o Fred hangs around with white swans
  - o Most swans are white
  - o Hypothesis: Fred is a rare black swan

Abductive reasoning is used routinely by doctors to make a diagnosis based on test results or by jurors making a decision based on evidence. It is commonly used in science as "inference to the best explanation." For example, the existence of Neptune was abduced from the odd orbit of Uranus. The discovery of the electron was abduced from the deflection of cathode rays.[3]

We will use abductive reasoning to demonstrate both the power of Digital Consciousness, as well as its strong likelihood of being true. One way to think about this is through the use of Venn diagrams.

Remember Set Theory?

Set theory is a branch of mathematics that involves the logic behind and relationship between groupings of objects. Sets can be easily described using Venn diagrams, as shown in Figure 4.1:

Set A - Things that are red   Set B - Things that are vehicles

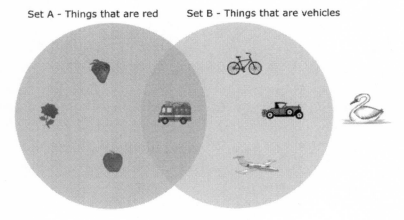

Figure 4.1

Imagine that Set A consists of all things that are red and Set B consists of all things that are vehicles. The *intersection* of Set A and B would therefore be all things that meet the criteria of both sets, namely vehicles that are red (for example, a fire engine). So in our diagram, there are four places that objects can be:

1. Set A only—such as a red rose
2. Set B only—such as a blue car
3. Both Set A and Set B; aka the *intersection* of Set A and Set B—such as a red fire engine
4. Neither Set A nor Set B—such as a white swan

We can also use sets and Venn diagrams to illustrate the abductive concept of the theory that best matches the data. Figure 4.2 shows how this would work:

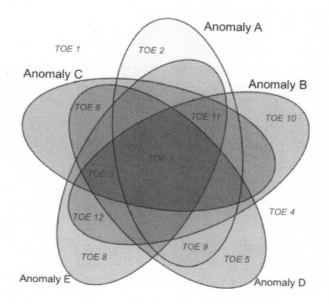

Figure 4.2

There are five anomalies in this "anomaly space" and 12 TOEs under consideration. TOEs 1 and 4 don't explain any of the anomalies at all, while TOEs 2, 5, 8 and 10 each explain only one. TOEs 6, 9 and 12 explain two anomalies, while TOEs 3 and 11 explain four of them (for example, TOE 11 explains all but Anomaly D). Only TOE 7 explains all five anomalies. Note that this doesn't imply that TOE 7 is "truth"—only that it provides an explanation for the five selected anomalies. All other TOEs can be rejected as true TOEs that provide a framework for everything; although many will still be called TOEs, because that is what people tend to call theories that work for everything in their particular field of interest.

Also, some TOEs might be able to "support" the anomalies but not have any real explanatory power. Often, a TOE (such as the "Grand Unified Theory" of physics) will provide a strong foundation for theories, but it is left to experimenters and theoretical scientists to come up with the specific theories that fill in the gaps.

In Figure 4.3, to take a real example, we analyze which theory of planetary motion explains various anomalies observed over the centuries. So, there is a set identified for each anomaly, which contains all of the theories that would explain that particular anomaly. Or, for another way of looking at it, theories are plotted and encircled by the anomalies that they satisfy. In this example, four anomalies are considered:

1. Retrograde motion: Retrograde motion is the apparent backward motion of the outer planets, which occurs as a speedier Earth passes them in its revolution around the sun.
2. Lack of stellar parallax: Stellar parallax would be the effect of seeing stars at different angles (relative to other celestial bodies) from the vantage point of different positions in orbit.
3. Phases of Venus change size: When Venus is between the Earth and the sun, it is at its largest apparent diameter, yet its phase is at its newest point (thinnest sliver). Whereas when it is opposite the sun, it is a full phase, but at its smallest diameter in the sky.
4. Stellar aberration: Stellar aberration is an annual variation in the apparent position of stars owing to the variance of

speed and direction of the Earth relative to the star.

Five different theories are summarized and included in the proper sets:

1. Aristotelian Geocentrism: Developed by Aristotle in the fourth century BCE, it is a simple model of the Earth at the center of the universe, with the sun, stars and planets revolving around it.
2. Ptolemaic Geocentrism: Claudius Ptolemaeus first formalized a comprehensive geocentric model of the universe that took into account retrograde motions of planets via "epicycles."
3. Copernican Heliocentrism with nearby stars: Copernicus' initial sun-centered model of the heavens still assumed that stars were fixed in a celestial sphere that wasn't that distant in comparison to the outer planets.
4. Tychonic Geo-heliocentrism with nearby stars: Tycho Brahe's model had planets revolving about the sun with the sun revolving about the Earth. It was highly equivalent to the pure Copernican heliocentric system from an observational standpoint.
5. Modern Keplarian Heliocentrism: Keplar's careful calculations of planetary motion allowed him to develop the more advanced laws of elliptical motion. That, coupled with advanced measurements showing that stars are very distant in comparison to the planets, brings us to the modern view of planetary motion.

Looking at Figure 4.3, it can be seen that one theory, Aristotelian Geocentrism, failed to explain three of the four anomalies (mostly likely because it was never fully formalized and based on detailed observation). Ptolemaic Geocentrism explained two of the anomalies—planetary retrograde motion and an apparent

47

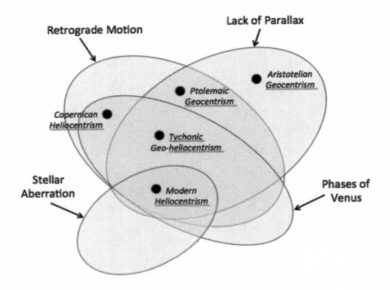

Figure 4.3

lack of stellar parallax—while early Copernican Heliocentrism more easily explained retrograde motion, but failed to explain the lack of stellar parallax, assuming that stars were relatively close to the sun. However, the heliocentric model very thoroughly explained the fact that the phases of Venus consistently vary with its apparent size in the sky. Tycho Brahe's combined geo-heliocentric model succeeded in explaining three of our chosen anomalies, with the added advantage of keeping the Church happy by placing the Earth at the center of the universe. By allowing for the possibility that stars were extremely distant compared to planets, the Copernican Heliocentric model also managed to explain three anomalies. However, a fourth anomaly was discovered in the 18th century, called Stellar Aberration. There was no way for even Brahe's model to explain that without Earth motion relative to stars, and so the only model capable of explaining all four anomalies is the Modern Heliocentric model that we have today.

There is another way to illustrate this. Instead of plotting "theory space," we could plot "anomaly space" (a set of anomalies) and use Venn diagrams to create sets of anomalies that are satisfied by particular theories. Figure 4.4 demonstrates this method. In this case, rather than seeking the theory that is at the intersection of all sets of anomalies, we look for the theory that supports all anomalies.

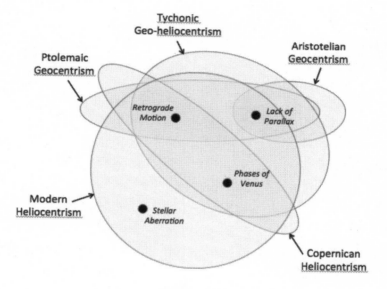

Figure 4.4

Aristotelian Geocentrism only encircles the "Lack of Parallax" anomaly. Ptolemaic Geocentrism covers Retrograde Motion plus Lack of Parallax, while Copernican Heliocentrism covers Retrograde Motion and Phases of Venus. Tychonic Geo-heliocentrism covers three of the four anomalies, but only Modern Heliocentrism covers all four. It's the same information, just presented in a different manner.

Of course, this is a contrived example, used for illustrative purposes only. Since the time frame for the development of these celestial motion theories was spread over 2,000 years and the various anomalies were observed over a similar period, there was never a point where anyone had five theories to apply to four anomalies at the same time. However, if they did, it would be clear to see what the winner was. I intend to use the same process to show that the Digital Consciousness model should be taken very seriously as a best fit to all of today's anomalies.

Chapter 5

Evidence—Digital

"Every physical quantity, every it, derives its ultimate significance from bits, binary yes-or-no indicators."
- *Physicist John Wheeler*

This is where the fun begins—compiling evidence for DCT. This chapter presents the evidence that our reality is ultimately discrete and digital deep down, while the next chapter presents the evidence for a consciousness-driven reality.

Which Came First, the Digital Chicken or the Digital Philosophy Egg?

As many scientists, mathematicians, futurists and philosophers are now embracing the idea that our reality is digital, it would be perfectly understandable to wonder if digital philosophy itself is tainted owing to the tendency of humans to view ideas through the lens of their times. We live in a digital age, surrounded by computers, the Internet and smart phones, and so might we not be guilty of imagining that the world behaves just as a multiplayer video game does? We probably wouldn't have had such ideas 50 years ago when, at a macroscopic level at least, everything with which we interacted appeared analog and continuous.

Actually, the concepts of binary and digital are not at all new. The I Ching is an ancient Chinese text that dates to 1150 BCE. In it are 64 combinations of eight trigrams (aka the Bagua), each of which clearly contain the first three bits of a binary code. Many other cultures, including the Mangareva in Polynesia (1450) and Indian (5th to 2nd century BCE), have used binary encodings for communication for thousands of years. Over 12,000 years ago,

African tribes developed a binary divination system called Odu Ifa.[1]

"There are 10 types of people in the world: those who understand binary, and those who don't."
- *Ian Stewart*

German mathematician and philosopher Gottfried Leibniz is generally credited as developing the modern binary number system in 1679, based on zeros and ones. Naturally, all of these other cultures are ignored, so that we can maintain the illusion that all great philosophical and mathematical thought originated in the EU. Regardless of Eurocentric biases, it is clear that binary encoding is not a new concept. But what about applying it to the fundamental construct of reality?

It turns out that while modern digital physics or digital philosophy references are replete with sources that only date to the mid 20th century, the ancient Greeks (namely Plato) believed that reality was discrete. Atoms were considered to be discrete and fundamental components of reality.

A quick clarification of the terms "discrete," "digital," "binary," "analog," and "continuous" is probably in order:

- Discrete: Having distinct points of measurement in the time or spatial domains.
- Digital: Having properties that can be encoded into bits.
- Binary: Encoding that is done with only two digits, zeros and ones.
- Analog: Having continuously variable properties.
- Continuous: The time or spatial domains are continuous.

So, to illustrate the differences between these properties, if we encode the value of some property (for example, length or voltage) digitally using 3 values (0, 1, 2), that would be digital, but not binary (rather, ternary). If we say that between any two points in time there is an infinitely divisible time element, but for each point the value of the measurement being performed on some property is represented by bits, then we would have a continuous yet digital system. Conversely, if time can be broken into chunks, such that at a fine enough temporal granularity there is no concept of time between two adjacent points in time, but at each of these time points the value of the measurement being performed is continuously variable, then we would have a discrete yet analog system.

Discrete or Continuous?

If reality is not discrete but continuous, this implies that there is infinite resolution in the time domain. That is, no matter how finely you examine time, there will always be an infinite continuum of points of time between any arbitrarily close instants in time. This isn't just a measurement challenge, but also a theoretical one.

When we watch a standard-definition movie in the theater, the movie is actually consisting of only 24 unique frames per second and yet we perceive continuous motion on the screen. This is because our brains can really only process 10 to 12 images per second.[2] Hence, the human eye and brain are simply not capable of noticing even the coarsest difference between discrete and continuous motion.

A slow-motion video camera, however, records images at a much higher frame rate, so that they can be played back at a normal rate, thereby slowing down the motion of the subject. Even slow-motion, instant replays in sports might not use frame rates exceeding 120 fps (frames per second) or, at most, 1,000 fps. Currently, the fastest frame rate in a video camera is 4.4 trillion images/second. Developed by scientists from the University of Tokyo and Keio University, and using a technique called femto-photography, the camera is able to capture chemical reactions.[3] This means that we can visually capture images of events in time that are separated by as little as 2.27E-13 seconds or .227 picoseconds. The shortest measured period of time ever is 1.2E-19 seconds or 12 attoseconds.[4]

Still, for all we know, there is plenty of action that goes on in between two instants of time separated by that amount. For example we know, theoretically, of half-lives (the average time to decay) of certain particles in the 1E-24 seconds range. We can also measure frequencies of gamma rays in the 300 EHz range, which means that the period of its waveform is 3.33E-21 or 3.33 zeptoseconds. So, without being able to directly observe it, we can infer that activity in our reality takes place at these tiny timescales. The question is: How far down does it go? Is there a limit to continuous time? At some point, is it discrete, like the frames in a movie?

The same assertions and questions can be made for the spatial dimensions. At its deepest level, is space continuous or discrete?

Whether or not space and time are discrete or continuous is actually pretty much the same question. Einstein showed that space and time are inextricably linked—effectively united by the speed of light. It would make little sense for one to be discrete and not the other. Reality would still be plagued by infinities. This is because it would still take an infinite number of states to represent a discrete-temporal continuous-spatial system or a continuous-temporal discrete-spatial system. Infinity times a finite number is still infinity.

To further clarify, imagine a reality that chunks along discretely in time, but has continuous spatial resolution. That spatial resolution is largely wasted, because between two adjacent points in time, an infinite number of spatial configurations *could* exist, but none *would* exist. Similarly, imagine a reality that is grainy spatially, but has continuous temporal resolution. That temporal resolution would be largely wasted, because between any two points in time that permit a spatial change (according to the relativistic limitation of the speed of light), there would be an infinite number of points in time in between those states, where the spatial configuration stays the same.

So, the question boils down to: Are time and space both discrete or both continuous?

Many physicists, like Julian Barbour and Rafael Sorkin, believe that time and space are discrete. Some identify the useful limit of spatial and temporal resolution with the Planck length of 1.6E-35 meters and the Planck time of 5.4E-44 seconds (also, in keeping with the idea that space and time are linked by the speed of light, this is the time that it takes for light to travel a Planck length). Physicists Ahmed Farag Ali, Mir Faizal and Barun Majumder, for example, have argued that "space and time exist only beyond a certain scale"; namely, the Planck scale.[5]

In reality, these are not hard theoretical limits, but represent theoretical limitations of measurement. That is, no matter what kind of measurement equipment one could conceivably dream up, they would be unable to measure any distance less than the Planck length, nor any time interval less than the Planck time, because quantum effects start to dominate the measurement.

Think about that for a moment. Quantum effects include things like uncertainty and the observer effect. Uncertainty is a statistical measure of randomness, based purely on information. The observer effect, as we shall see, is inextricably tied to consciousness. Thus, consciousness, information and the limitations of reality are completely tied together.

Oddly enough, Jacob Bekenstein showed that a Planck area (Planck length squared) is exactly how much a black hole's surface increases when it "swallows" one bit of information. This alone is a fascinating result and seems to imply that information is tied to the apparent limit of spatial resolution. This, of course, makes sense in a macroscopic context. Bits of data have to be stored in some sort of spatial orientation, whether they be magnetic domains on a hard disk, electrically doped segments of a semiconductor or proteins on a DNA strand. Double the amount of hard disk space, semiconductor space or DNA surface area and you can double the amount of information stored. Double the amount of units of Planck area and you can double the bits that encode reality. Again, data and limitations of reality appear in the same relationship.

In the realm of consciousness-driven digital philosophy, it is my contention that the evidence strongly supports reality being *discrete* and *digital*; that is, time moves on in "chunks" and at each discrete point in time, every property of everything can be perfectly represented digitally. There are no infinities.

I believe that this is a logical and fundamental conclusion, regardless of the fact that we live in a digital age. There are many reasons for this, which shall be presented later in this chapter.

But first, owing to the many upcoming references to aspects and consequences of quantum mechanics in this chapter and the next, a brief summary of the scientific theory that changed physics forever in the 20th century, and will likely change philosophy forever in the 21st century...

A Ridiculously Brief Primer on Quantum Mechanics

Trying to summarize quantum mechanics in a small section of a chapter of a book is like trying to fit an ocean into an eyedropper. And yet, without some overview of it, the discussions in this and the next chapter might be a little hard to follow.

> "Anyone who is not shocked by quantum theory has not understood it."
> - *Niels Bohr*

In a nutshell, quantum mechanics is the branch of physics that describes what happens at very small (atomic) scales. And, specifically, what happens at those scales is that matter and energies appear to be quantized.

Max Planck's model of blackbody radiation and Einstein's quantum theory of the photoelectric effect, along with the experimental results supporting these theories, established the foundation for quantum mechanics. Niels Bohr and Erwin Schrödinger developed equations that described what was going on at the atomic level. In their view, electrons would only "fit" into certain orbital levels around the nucleus of an atom just like the vibrations of guitar strings can only fit a certain set of waveforms owing to their length. Quantum effects had to do with how these discrete properties manifested in the observable world, such as specific energy levels of light generated as electrons shifted in orbital layers around the atom.

Within years of the beginning of this nascent field of study, some very mysterious properties were emerging from a combination of experimental results and theory. In the double-slit experiment, for example, light was shown to exhibit either properties of a wave or properties of a particle, depending on the type of measurement performed. This became known as the wave-particle duality nature of light and was ultimately extended to all forms of matter (particles, atoms, molecules). The "superposition principle" evolved from Wolfgang Pauli's "exclusion principle," which states that two particles in the same system cannot be in the same quantum state simultaneously. In superposition, however, the particles can exist simultaneously in multiple states and the actual state is not determined until a measurement is made. Since this immediately affects the resulting state of the other particle, such particles were said to be entangled. Einstein, Boris Podolsky and Nathan Rosen wrote a famous paper in 1935, in which they presented what is now known as the EPR Paradox (EPR for their last names). This paradox asserts that if one takes two entangled particles and separates them arbitrarily far, and then makes a measurement on one, the state of the other would be immediately determined, seemingly violating the law of relativity that no information can travel faster than light.

Many different interpretations of quantum mechanics have been presented over the years to account for these anomalies and paradoxes. The standard and original interpretation, developed by Niels Bohr and Werner Heisenberg, is referred to as the Copenhagen interpretation, owing to the location where the two worked during the formulation of this theory. In the Copenhagen interpretation, particles are described by a wave function, which essentially represents the likelihood that a particular property of a particle (for example, position) will be measured as such. The act of measuring, or observing, results in what is referred to as the "collapse of the wave function" into definitive properties.

In the 1950s, physicist Hugh Everett developed a radically different interpretation of quantum mechanics that completely avoids the messiness of wave function collapse and replaces it instead with the messiness of parallel universes. In his theory, every time that a probabilistic decision is made at the quantum level (does the photon go through this slit or that slit, or does the radioactive particle decay, thereby killing Schrödinger's infamous hypothetical cat?), what actually happens is that the universe splits in two. In one universe, the particle doesn't decay and the cat stays alive. In the other universe, the particle decays and the cat dies.

The reason that the cat is alive when you open the box is because we happen to live in the universe in which the cat is alive. Since probabilistic decisions are made at the quantum level zillions of times per second, the Everett interpretation of quantum mechanics says that somewhere there are universes where you are a movie star, universes where you don't exist, universes where you read everything on this page until this exact word and then put down the book forever, universes with no life, and pretty much every other universe that you can possibly imagine.

Cosmologist Max Tegmark estimates that there should be 10 to the power of 10^{118} total universes, which is impossible to imagine.[6] Perhaps even stranger, it turns out that most physicists actually believe this! In 1995, political scientist L. David Raub conducted a poll of 72 of the "leading cosmologists and other quantum field theorists," and found that 58 percent believed that the Many Worlds Interpretation (MWI) was true, including Stephen Hawking and Nobel Laureates Murray Gell-Mann and Richard Feynman. Another 13 percent believed that it was possible but that they weren't yet convinced.[7]

So where are these universes, anyway, and can we jump from one to the other? As you might expect, they are not nearby or even part of our normal three-dimensional space, but rather exist in an abstract infinite-dimensional space called the Hilbert Space. And no, we can't get there from here, so there goes your bright idea on winning the lottery.

Many other varieties of quantum mechanics interpretations have been developed over the years, in order to get around some of the strangeness that accompanies the others...

Elvidge's Postulate of Countable Interpretations of QM: *The number of interpretations of Quantum Mechanics always exceeds the number of physicists who ponder Quantum Mechanics.*

Let's count the various "interpretations" of quantum mechanics:[8]

1. Bohm (aka Causal or Pilot-wave)
2. Copenhagen
3. Cosmological
4. Ensemble
5. Ghirardi-Rimini-Weber
6. Hidden measurements
7. Many-minds
8. Many-worlds (aka Everett)

9. Penrose
10. Possibilist Transactional (PTI)
11. Relational (RQM)
12. Stochastic
13. Transactional (TIQM)
14. Von Neumann-Wigner
15. Digital Consciousness (DCI, aka Elvidge)

Unfortunately, you won't find the last one in Wikipedia. Give it about 30 years.

OK, armed with a background on quantum mechanics, how about some of that digital evidence...

Evidence: Breakdown of Infinite Resolution in Physics

"Two things are infinite. The universe and human stupidity. And I'm not so sure about the universe."
- Albert Einstein

According to science philosophy professor Amit Hagar, "if you look around in physics, actual infinity is banned, singularities are avoided, and divergences are tamed."[9]

One example of what Hagar is referring to is the so-called trans-Planckian problem, which is the consideration of what would happen at scales below the Planck level, given the physics that we have today. Basically, it all breaks down. One instance of this breakdown is Hawking radiation, which is the radiation that emanates from black holes owing to quantum effects near the black hole event horizon. Theoretically, Hawking radiation should include arbitrarily large frequencies, including those above the Planck frequency of 1/(Planck length). But such frequencies are not observed. So, either physics changes at the Planck level or nothing exists below it.[10] What this means is that there is not infinite resolution, which means that space and time are discrete or quantized.

Infinite resolution would imply that relativity and quantum mechanics can't coexist, at least with the best physics that we have today. Our favorite contenders for rationalizing relativity and quantum mechanics are string theory and loop quantum gravity. And they only work with minimal length (aka discrete) scales.[11, 12, 13]

Furthermore, we actually observe discrete behavior in subatomic physics. For example, a particle's spin value is always quantized; there are no intermediate states. Fermions have spins designated in half integers, such as +/-½, whereas bosons have spins designated in integers. There is no such thing (it has never been observed), for example, as a spin of ¼ or .237. So, if an electron has a spin value of ½ and a magnetic field changes it to -½, the transition is immediate. It does not go through a gradual transition of spin numbers. Such behavior is completely anomalous in continuous space-time.

Evidence: Origin Logic

"If it turns out that there is a God, I don't think that he's evil. But the worst that you can say about him is that basically he's an underachiever."
- *Woody Allen*

Let's break down the possibilities of our reality, in terms of origin and behavior:

1. Our reality was created by some conscious entity and has been following the original rules established by that entity. Of course, we could spend a lifetime defining "conscious" or "entity," but let's try to keep it simple. This scenario could include traditional religious origin theories (for example, God created the heavens and the Earth). It could

also include the common simulation scenarios, à la Nick Bostrom's "Simulation Argument."[14]

2. Our reality was originally created by some conscious entity and has been evolving according to some sort of fundamental evolutionary law ever since.

3. Our reality was not created by some conscious entity, and its existence sprang out of nothing and has been following primordial rules of physics ever since. To explain the fact that our universe is incredibly finely tuned for matter and life, materialist cosmologists dreamed up the idea that we must exist in an infinite set of parallel universes, and via the anthropic principle, the one we live only appears finely tuned because it has to in order for us to be in it. Occam would be turning over in his grave.

4. Our reality was not created by some particular conscious entity, but rather has been evolving according to some sort of fundamental evolutionary law from the very beginning.

I would argue that in the first two cases, reality would have to be digital. For, if a conscious entity is going to create a world for us to live in and experience, that conscious entity is clearly highly evolved compared to us. And, being so evolved, it would certainly make use of the most efficient means to create a reality. A continuous reality is not only inefficient, but it is also theoretically impossible to create, because it involves infinities in the temporal domain as well as any spatial domain or property. Therefore, it demands infinite resources.

I would also argue that in the fourth case, reality would have to be digital for similar reasons. Even without a conscious entity as a creator, the fundamental evolutionary law would certainly favor a perfectly functional reality that doesn't require infinite resources.

Only in the third case would there be any possibility of a continuous analog reality. Even then, it is not required. Hence, there is no reason to assume, a priori, that the world is continuous.

As mentioned before, from the standpoint of simplicity of argument, Occam's razor, option 3, should be ruled out, leaving only three other categories of origin, each of which strongly favor being digital.

For many other reasons, as will be clear from the evidence compiled in this book, I tend to favor reality option 4. No other type of reality structure and origin can be shown to be anywhere near as consistent with all of the evidence (philosophical, cosmological, mathematical, metaphysical and experimental). And it has nothing to do with MMORPGs or the smart phone in my pocket.

Evidence: The Simulation Argument

[Note: this section is a refinement of the corresponding section from The Universe—Solved! *and has been updated to incorporate new understandings and ideas.]*

Nick Bostrom, Director of the Oxford Future of Humanity Institute and Professor in the Faculty of Philosophy at Oxford University, asks the following question in his paper: "Are you Living in a Computer Simulation?": "If there were a substantial chance that our civilization will ever get to the posthuman stage and run many ancestor-simulations, then how come you are not living in such a simulation?" By "posthuman," he refers to an evolutionary stage, whereby humans have mastered most forms of technology that are consistent with the laws of physics.

He argues, "at least one of the following is true:

1. The human species is very likely to go extinct before reaching a 'posthuman' stage;
2. Any posthuman civilization is extremely unlikely to run a significant number of simulations of their evolutionary

history (or variations thereof);

3. We are almost certainly living in a computer simulation. It follows that the belief that there is a significant chance that we will one day become posthumans who run ancestor-simulations is false, unless we are currently living in a simulation."[15]

The first consideration is the question of when it would be possible to generate a reality via a computer simulation that is consistent with our observations. Futurist (and self-proclaimed "Transcendent Man") Ray Kurzweil argued in his 2005 book *The Singularity is Near* that simply following Moore's law, it will be likely that we will have the technology to emulate our reality to a resolution equivalent to the observable granularity that we have defined here by the year 2020. While this may be true of resolution, there is much more to creating a fully immersive reality simulation than sufficient image resolution. The complexity of reality must approximate that which we are used to. In addition, there must be a technology supporting the suppression of existing memories (for example, saved to temporary storage and then erased from the brain, to be reinstated later after the completion of the sim), in order to believe fully in the reality of the presented simulation. Finally, there must be a delivery method that provides a 360-degree surround simulation with a frame rate exceeding our brain's refresh rate. Such a delivery method would most effectively be an implant that intercepts our incoming sensory cortex signals. While I suggest that 2050 might be a more realistic date for these capabilities, whether the time frame for this technology is 2020 or 2050 doesn't really matter, because it is a small interval on the scale of human evolution.

Considering Bostrom's second posthuman possibility, I would argue that given our technological history, it certainly seems unlikely that humans will ever be presented with a fascinating and feasible line of inquiry and not pursue it.

Consider the atomic bomb, cloning, genetic modifications, AI pursuits and nanotech, to name but a few. Does anyone really believe that the entire world would make a conscious effort not to continue to push the virtual reality envelope, especially considering the jump-start that the technology already has in the gaming industry? Therefore, Bostrom's second posthuman possibility is very unlikely, leaving only the first and the third. Essentially, that says that either we are currently living in non-simulated reality and will fail to reach posthuman status or we have already reached it and are living in a simulation.

Is there any way to determine which is more likely? Possibly, using the following timeline argument...

Imagine a 100,000-year timeline of human evolution (admittedly, this is somewhat arbitrary—conventional science has protohumans at 7 million years ago, abstract thinking at 50,000 years ago,[16] development of civilization at 6,000 years ago)—50,000 years into the past and 50,000 years into the future. Nick Bostrom's posthuman phase stretches all the way from 2050 to the year 50000. Our apparent reality is the year 2017. We determined earlier that we are likely to be either in the apparent reality of 2017 or living in a simulation that was kicked off in the posthuman era of the timeline, and that there is no way to tell the difference. Since the posthuman era is about 48,000 years long, it seems more likely that we are somewhere in that range, rather than merely 33 years prior to that range.

Statistically, one might come up with the probabilities by considering the odds of picking a random point on the timeline by, say, throwing a dart at the timeline. We neglect the period prior to 2017, because we know we can't be living in that period. The probability that the dart lands in the current era (2017–2050) is 0.0009. The probability that the dart lands in the posthuman era (2050–50000) is .9991. So, by this argument, we are almost certainly living in a simulation. Again, you sharp-minded readers are no doubt thinking, this is entirely dependent on the choice of 50000, which is fairly arbitrary.

It came from assuming that we might be in the middle of an evolutionary epoch that began when traditional science says that humans developed abstract thought. One might argue that the timeline should really end at the point that our civilization is likely to end. For the pessimists, that would be sooner rather than later, since they would believe that we will never make it to a Type I Kardashev civilization (defined by astronomer Nikolai Kardashev as the point at which a civilization learns to harness all of its planet's energy), let alone 50000. What about 2029, when asteroid 2004 MN4 does its flyby of planet Earth? Or 2025, when we might create an AI that outsmarts us (the "Terminator" scenario)? Or possibly 2045, per the technical singularity argument? Or 2191, when Moore's Law allows us to hit the Planck energies?

The pessimist's options are endless. Even by averaging out the common doomsday scenarios, we stand roughly a 50 percent chance of living in a simulation right now. If you take the optimistic point of view and assume that we are capable of making it past the Type I Kardashev transition, the odds are good that we would continue on for millions of years, in which case the probabilities of our reality being a simulation are very nearly 100 percent.

It is clear how this relates to Digital Consciousness. If it is likely that we are living in a simulation, then by definition we are living in a Digital Consciousness reality. However, it is also possible that we are living in a Digital Consciousness reality that is *not* a simulation, in the traditional sense in which that word is used, as in Origin Logic option 4.

Evidence: Matter as Data

Extrapolation is a technique for projecting a trend into the future. It has been used liberally by economists, futurists and other assorted big thinkers for many years, to project population growth, food supply, market trends, singularities, technology directions, skirt lengths and other important trends. It goes something like this:

If a city's population has been growing linearly by 10 percent per year for many years, one can safely predict that it will be around 10 percent higher next year, 21 percent higher in two years and so on. Or, if chip density has been increasing by a factor of 2 every 2 years (as it has for the past 40), one can predict that it will be 8 times greater than today in 3 years (known as Moore's Law). Ray Kurzweil and other singularity fans extrapolate technology trends to conclude that our world as we know it will come to an end in 2045 in the form of a technological singularity. Of course, there are always unknown and unexpected events that can cause these predictions to be too low or too high, but given the information that is known at the current time, this is still a useful technique.

To my knowledge, extrapolation has not really been applied to the problem that I am about to present, but I see no reason why it couldn't give an interesting projection for the nature of matter.

In ancient Greece, Democritus put forth the idea that solid objects were comprised of atoms of that element or material, either jammed tightly together, as in the case of a solid object, or separated by a void (space). These atoms were thought to be little indivisible billiard-ball-like objects made of some sort of "stuff." Thinking this through, it was apparent that if atoms were thought to be spherical and they were crammed together in an optimal fashion, then matter was essentially 74 percent of the space that it takes up, the rest being empty space. So, for example, a solid bar of gold was really only 74 percent gold "stuff," at most.

That view of matter was resurrected by John Dalton in the

early 1800s and revised once J. J. Thomson discovered electrons. At that point, atoms were thought to look like plum pudding, with electrons embedded in the proton pudding. Still, the density of "stuff" didn't change, at least until the early 1900s, when Ernest Rutherford determined that atoms were actually composed of a tiny dense nucleus and a shell of electrons. Further measurements revealed that these subatomic particles (protons, electrons and later, neutrons) were actually very tiny compared to the overall atom and, in fact, most of the atom was empty space. That model, coupled with a realization that atoms in a solid actually had to have some distance between them, completely changed our view on how dense matter was. It turned out that in our gold bar only 1 part in 10^{15} (1,000,000,000,000,000) was "stuff."

That was until the mid 1960s, when quark theory was proposed, which said that protons and neutrons were actually comprised of three quarks each. As the theory (aka Quantum chromodynamics or QCD) is now fairly accepted and some measurement estimates have been made of quark sizes, one can calculate that since quarks are between 1,000 and 1 million times smaller in diameter than the subatomic particles that they make up, matter is now 10^9 to 10^{18} times more tenuous than previously thought. Hence, our gold bar is now only about 1 part in 10^{30} (give or take a few orders of magnitude) "stuff" and the rest is empty space. By way of comparison, about $1.3*10^{32}$ grains of sand would fit inside the Earth. So matter is roughly as dense with "stuff" as one grain of sand is to our entire planet.

So now we have three data points to start our extrapolation. Since the percentage of "stuff" that matter is made of is shrinking exponentially over time, we can't plot our trend in normal scales, but need to use log-log scales.

And now, of course, for a possible fourth data point, we have string theory, which says that all subatomic particles are really just pieces of string vibrating at specific frequencies, each string possibly having a width of the Planck length. If so, that would make subatomic particles all but 10^{-38} empty space, leaving our gold bar with just 1 part in 10^{52} of "stuff."

Gets kind of ridiculous, doesn't it? It isn't hard to see where this is headed.

In Figure 5.1, I plotted the density of matter according to science at various points over the years. For the string theory data point, I picked the year 2033 because, given Moore's Law and the fact that it would take energies 10^{14} greater than are currently being used at the Linear Hadron Collider (LHC) at CERN, 2033 might be about the year that we could validate or invalidate string theory. I predict that, by then, if string theory is not validated, there will be some other theory of matter that has an approximately equivalent density.

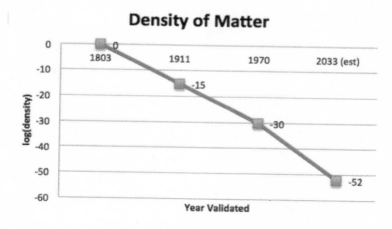

Figure 5.1

Another way to think about this trend is to recognize that it is the experimental energies available to us that have made it possible to probe the depths of matter. The greater the experimental energy, the deeper the probe and the more tenuous matter appears to be. Why would we expect an end to this trend? Wouldn't it make sense that higher energies will result in decomposing the previously smallest units of matter to the effect of matter losing another dozen or so magnitudes of tenuousness?

In fact, if particles are comprised of strings, why do we even need the idea of "stuff?" Isn't it enough to define the different types of matter by the single number—the frequency at which the string vibrates?

What is matter, anyway? It is a number assigned to a property of an object that has to do with how that object behaves in a gravitational field. In other words, it is just a rule.

The fact is, we don't really experience matter. What we experience when we look at something is the electromagnetic radiation influenced by some object that we call matter. When we touch matter, we are experiencing the effect of the laws of electromagnetic force owing to the repulsion of charges between the electron shells of the atoms in our fingers and the electron shells of the atoms in the object.

In any case, if you extrapolate our scientific progress, it is easy to see that the ratio of "stuff" to "space" is trending toward zero, as shown in Figure 5.1. Which means what?

That matter is most likely just data. And the forces that cause us to experience matter the way we do are just rules about how data interacts with itself.

And data is digital.

"As a man who has devoted his whole life to the most clear headed science, to the study of matter, I can tell you as a result of my research about atoms this much: There is no matter as such. All matter originates and exists only by virtue of a force which brings the particle of an atom to vibration and holds this most minute solar system of the atom together. We must assume behind this force the existence of a conscious and intelligent mind. This mind is the matrix of all matter."
- *Max Planck*

Evidence: Equations Creating Reality

Were our reality what it appears to be, solutions to equations should only make sense in the context of describing reality. However, sometimes it seems to be the other way around, where data and rules don't manifest from the reality but rather, they create the reality. To cite one of many examples, it has been shown that the negative frequency solutions to Maxwell's equations actually reveal themselves in components of light.[17] This is extremely curious, because in algebra we learn that there are sometimes solutions to equations that don't make sense in the "real" world.

For example, let's solve the simple mathematical problem of calculating how much fence is needed to enclose a square area of 9 square meters. The solution is simple: one side of the square is $\sqrt{9}$ and so the length of fence is $4*\sqrt{9}$. However, there are two solutions to $\sqrt{9}$: 3 and -3. Clearly, it doesn't make sense to have -3 meters of length of fence, so the -3 solution to the equation is thrown out.

An analogous situation is found in the solutions to Maxwell's equations, shown below in Figure 5.2. Since these equations are also in quadratic form, they have negative frequency solutions. Those negative frequencies don't correspond to anything in the real world and so they would normally be ignored. However, it turns out that in some scenarios the light field associated with the negative frequencies themselves also have positive and negative frequency solutions.

In 2012, for the first time, those "secondary" positive frequencies were observed in experiments. In a "real world" view of reality, where equations describe a reality that already exists, this shouldn't happen, because nothing should be generated from a negative frequency. However, in a world where reality is the manifested *result* of a process driven by an equation, such artifacts *should* exist. Since we see the latter, it lends support to the idea that equations drive or *create* reality as they would in a program.

$$\left(c^2\nabla^2 - \frac{\partial^2}{\partial t^2}\right)\mathbf{E} = 0$$
$$\left(c^2\nabla^2 - \frac{\partial^2}{\partial t^2}\right)\mathbf{B} = 0 \quad c = \frac{1}{\sqrt{\mu_0\varepsilon_0}}$$

Figure 5.2 Electromagnetic wave equation[18]

Another recent and pretty spectacular example of equations found in nature was discovered by physicist James Gates. Supersymmetry theory in physics (aka SUSY) describes the relationships between different particles as well as the interactions between particles. When depicting SUSY physics relationships graphically, the results are specific diagrams or patterns. Because of the similarity of these patterns to West African Ashanti symbols called adinkras, they have been referred to as such.[19]

In systems that involve the transmission of data through some communication channel, the data, in the form of 1s and 0s, are encoded in some manner (for example data-encoded radio waves through the air, or a fiber optic cable, or satellite telemetry). During the transmission from one point to another, there is always a chance that the data can be corrupted; a bit flipped. In some cases, it doesn't really matter that much. For example, a video being transmitted could undergo many flipped bits and the end user would never know it. However, for a data file that contains bank balances, the flipping of a bit can be catastrophic.

For example, if 100,010,100,101,000,000,001 represents a transaction from your Swiss bank account in dollars, it would correspond to a modest (by Swiss bank account standards) $1,133,057 or a little more than a cool million. However, if that first bit got flipped in the communications channel, the amount would become $84,481 and somebody would have a lot of explaining to do at the Bugatti dealer.

Fortunately, some bright mathematicians like Richard Hamming and Claude Shannon developed some things called error-correcting codes which, when sent along with the data, allow the receiving end to correct for flipped bits. There is a category of such codes called—I kid you not—"doubly even selfdual linear binary error-correcting block codes." So what does this have to do with SUSY and adinkras? Just that Gates and his colleagues found that such error-correcting codes appear in certain adinkras that describe a certain transformation of SUSY relationships. The upshot is that error-correction codes may occur in nature. As Gates himself observes, "let us imagine that this alternative Matrix-style world contains some theoretical physicists and that one of them asks, 'How could we discover whether we live inside a Matrix?' One answer might be 'Try to detect the presence of codes in the laws that describe physics.'"[20]

Evidence: Bert & Ernie and the Granularity of Space and Time

GEO600 is a large gravitational wave detector located in Hanover, Germany. Designed to be extremely sensitive to fluctuations in gravity, its purpose is to detect gravitational waves from distant cosmic events. In the late 2000s, however, it had been plagued by inexplicable noise or graininess in its measurement results. Craig Hogan, director of Fermilab's Center for Particle Astrophysics, considered the possibility that the instrument had reached the limits of space-time resolution and that this might be proof that we live in a digital hologram. Using physicists Leonard Susskind and Gerard 't Hooft's theory that our 3D reality may be a projection of processes encoded on the 2D surface of the boundary of the universe, he pointed out that, like a common hologram, the graininess of our projection may be at much larger scales than the Planck length (10^{-35} meters), such as 10^{-16} meters. In his research paper coauthored with Mark Jackson, "Holographic Geometry and Noise in Matrix Theory,"[21] he claimed, "the holographic noise derived here has previously been shown to approximately agree with the spectrum of currently unexplained continuum noise in the best operating interferometer, GEO600."

"If the GEO600 result is what I suspect it is, then we are all living in a giant cosmic hologram," added Hogan.[22] Driven to find the answer, his team built the world's most sensitive laser interferometer, dubbed a Holometer, at Fermilab. After collecting data for a year, Hogan and his team announced late in 2015 that the results appeared to indicate that space-time is not discrete at Planck scales. However, it does not mean that reality isn't digital—only that if it is, it must be digital at an even finer granularity.

As soon as one door closes another opens, almost as if the universe *wants* to be understood. Some anomalously high-energy particles from space were discovered in 2013 and 2014 at the IceCube Neutrino Observatory in Antarctica. The first few were named after *Sesame Street's* Bert, Ernie and Big Bird, and dozens more have been discovered since. What has physicists puzzled, however, isn't the high-energy neutrinos, but the lack of even higher energy neutrinos. According to Nobel Prize winner Sheldon Glashow, standard models of physics predict higher energies of neutrinos in some nuclear reactions; yet there seems to be an upper bound to the neutrino energies observed.

One theory proposed by University of Hawaii physicist John Learned is that these "super-short wavelengths of the neutrinos are sampling the structure of space and there will not be higher energy neutrinos seen." It is as if the energy of the particle is limited by the spatial gridwork through which it must travel.[23]

Again, such theories are highly speculative, but it does seem that as we probe deeper and deeper into the fabric of space-time, we begin to see anomalies that, at a minimum, call for a reanalysis of known laws of physics at small scales, if not provide evidence for the digital nature of space itself.

Evidence: Similarity of Information Systems to Reality

The very nature of the computational mechanisms of a computer is essentially the same as quantum mechanics—a sequence of states, with nothing existing or happening between the states. The resolution of any program is analogous to the spatial resolution of our reality, just at a different level. In fact, carrying Moore's law (which has been consistent over the past 40 years) forward, computers will reach the Planck resolution in 2192.

However, it is not necessary to model reality all the way to that level for the model to be indistinguishable from our reality. Only the *observed* elements of reality need to be modeled and then only down to a resolution that matches the observational limitations of our measurement devices. A program can do this dynamically. Therefore, given Moore's law and the limitations of "observational reality," we should be able to create virtual realities that are indistinguishable from our current reality within 20 years or so. The very fact that our reality appears to be quantized may be considered strong evidence that it is information based.

Many researchers have also noted that a simulation model solves the "prime mover" philosophical problem. While the Big Bang Theory implies a universe that arises from nothing, which has no grounding in objective reality, a virtual reality can easily "boot up" from an external context. In information systems researcher Dr. Brian Whitworth's papers, "The emergence of the physical world from information processing"[24] and "Quantum Realism,"[25] he outlines several additional examples of circumstantial evidence that our universe is a simulation, including the apparent randomness.

There appears to be no hidden variables that deterministically cause the decay of a radioactive atom, for example. Instead, atoms decay according to the rules of half-life probability. Each radioactive isotope has its own characteristic half-life, which is defined as the amount of time that has to elapse before half of the atoms of that isotope have decayed into something else. So, for instance, uranium-235 has a half-life of 700 million years. That means that there is a 50 percent probability that a given U-235 atom will decay within 700 million years. That's all that can be said about it. There appears to be nothing else that influences the point in time when the decay occurs; rather, it is purely probabilistic. It never made sense to Einstein (who, famously, said "God does not play dice") nor to many other scientists that particles would behave randomly as opposed to following deterministic rules. However, in a computational model, random number generators are a simple concept even in today's systems and can easily be used to drive apparently random behavior. Extending this argument, the quantum "probability wave" has no basis in objective reality, but is easily created computationally.

In his paper, "Exploring the virtual reality conjecture," Whitworth identifies "ten reasons to suspect that the physical world is a simulation."[26] Some of his other more convincing arguments include:

- *The speed of light.* An objective reality has no reason for a maximum speed, but every simulation screen has a maximum refresh rate that limits local transfers.
- *Superposition.* Objective entities cannot spin in two directions at once as quantum entities do, but a program can divide itself to do this.
- *Quantum tunneling.* An electron "tunneling" through an impenetrable field barrier, like a coin popping out of a perfectly sealed glass bottle, is impossible if objects con-

tinuously exist, but not if they are discrete event frames.

Evidence: Lack of Evidence for Continuous Reality

It may be argued that although there is specific evidence for a digital reality, if there is an equivalent or greater set of evidence for a continuous reality, then the scale would tip in the direction of that possibility. So how do the continuous reality arguments stack up?

Not very well.

David Tong, professor of theoretical physics at the University of Cambridge, developed perhaps some of the most compelling arguments in favor of continuous reality. In fact, he took second place prize in the Foundational Questions Institute's essay contest: "Is Reality Digital or Analog?"[27] However, while his arguments were creative, they were not conclusive in the least.

Tong points out that the "quantum" in quantum mechanics is a discrete characteristic that emerges from continuous properties. In this case, he is correct, at least as far as our current understanding goes. It is analogous to harmonics and resonances. A guitar string, when plucked, vibrates at a certain frequency or tone. Effectively the properties of the string, such as its length, mass and tension, define the fundamental frequency at which it vibrates. This vibration sets up what is known as standing waves on the string, which are called such because they appear to stand still. A finite number of standing wave wavelengths will exist on the string and this number again is a function of the properties of the string. So, a discrete characteristic (the finite number of standing waves) emerges from the "more" continuous motion of the vibrating string. I say "more" because that motion is not at all fundamental; it is just what we observe.

Schrödinger's equation (developed by physicist Erwin Schrödinger in 1925), he points out, defines the nature of the quanta that emerge from excited hydrogen atoms. While

Schrödinger's equation worked well for the very simple case of an atom consisting of a single electron, it was insufficient for even the second simplest atom, helium. Physicists have since given up trying to create a formula that predicts quanta energies from atoms beyond hydrogen. So, in that specific case, Tong is correct that a discrete characteristic can emerge from an underlying "more" continuous process.

The flaw in his argument, however, is in making the leap that reality is continuous "all the way down." Nothing in his example logically leads to such a conclusion. In fact, it is easy to demonstrate that there are layers of alternating discreteness and continuousness in our reality. A case in point is the set of waves in a pond. The waves are discrete, on the order of meters in size, but a deeper reality is the water in the pond, which appears to be continuous. However, that water is discrete at a deeper layer of reality, in that it is composed of a countable number of atoms and, in turn, subatomic particles, on the order of 10^{-10} meters. However, deeper still are the continuous fields that supposedly generate the particles themselves. And yet, much deeper (20 orders of magnitude deeper) is the Planck realm, which appears to be discrete. Below that, nothing can be said, because there are neither theories nor experimental data to describe that level. So, at the deepest layer we know about, things still appear discrete.

Tong's argument is way up at such a high level that everything could easily look continuous. What is important to our question is not any level other than the deepest possible one. And one cannot infer anything about the true nature of reality by looking at evidence that corresponds to anything other than the deepest level. Reality can't be said to be continuous because the fields that beget particles are continuous any more than reality can be said to be discrete because you can count the number of waves in the ripple of a pond.

Tong also references chiral fermions, which have to rotate 720 degrees to return to their original state. Arguing that such characteristics can't be explained computationally is fallacious. Watch:

```
switch (fermion type) {
    case chiral:
        switch (rotation) {
            case 360:
                ...
                break;
            case 720:
                fermion (next state) == fermion
(last state);
                break;
            case other:
                ...
                break;
        }
        break;
case other:
        switch (rotation) {
            case 360:
                fermion (next state) == fermion
(last state);
                break;
            case other:
                ...
                break;
        }
        break;
}
```

It isn't difficult to shoot similar holes in any other argument for continuous reality. As renowned MIT cosmologist Max Tegmark says: "We've never measured anything in physics to more than about sixteen significant digits, and no experiment has been carried out whose outcome depends on the hypothesis that a true continuum exists, or hinges on nature computing something uncomputable."

Summary

The evidence that reality is digital is almost overwhelming. The other option, continuous reality, is practically impossible. There is no evidence to support it, save for the macroscopic observation that things "seem" continuous. Just like a movie consisting of 24 discrete frames per second *seems* continuous. Or how the world *seems* flat at ground level. Subjective observations of reality are clearly not really evidence of its nature. One must look deeper.

And when you do, you realize that:

- Physics based on a continuous model of reality completely breaks down at sufficiently small scales, implying that only discrete models of reality can be real. Loop quantum gravity and string theory, the most advanced theories that we have of how the world works, are based on discrete space and time. Subatomic particles have discrete properties.
- Philosophically, only simulation-based theories of reality are consistent with the logic of technological evolution.
- Simulation-based theories of reality provide by far the best explanations for the observations of a finely tuned universe.
- Only digital realities are efficient thermodynamically and if our world evolved according to fundamental laws of evolution, it must therefore be digital.
- The process of scientific discovery is leading inexorably to the conclusion that matter is composed only of information.

- We are starting to learn how discrete mathematical equations appear to be responsible for some behaviors of reality. And we have recently discovered error-correction code algorithms in the equations of physics.
- Digital models of reality are the only models that provide explanations for the prime mover, the apparent randomness in physical processes, probability functions in quantum mechanics and rationales for the speed of light.

Chapter 6

Evidence—Consciousness

"There is nothing that we know more intimately than conscious experience, but there is nothing that is harder to explain."
- *Cognitive Scientist David Chalmers*

While the evidence that our reality is fundamentally digital is strong and comprehensive, as shown in the last chapter, the evidence that our reality is based on consciousness is perhaps even more definitive.

What does it mean to say that reality is consciousness-based? In short, it means that the subjective experience that we have is not an artifact of the complexity of a physical brain, but rather, the other way around—all that we experience is rooted in a consciousness that exists above and beyond the physical brain.

One of the best ways to understand this is to consider a dream. It is fairly obvious to everyone that the "reality" that you experience in a dream is generated by your consciousness, regardless of where the seat of that consciousness is. It is easy to accept that idea because dreams tend to be personal and individually generated.

However, there are exceptions to that rule, such as mutual lucid dreaming, where multiple people share a somewhat common dream experience. I say "somewhat" intentionally, because the experience may have common elements but not be identical. That tells you two very interesting things:

1. The fact that multiple people can share any aspect of a dream experience that is statistically significant beyond coincidence (as mutual lucid dreams generally are) means

that the origin of the "dream world" or "dream reality" cannot be entirely in the brain.

2. The fact that each person's dream experience is *not* identical indicates that they don't come entirely from the same source and that at least some aspects of the experience are purely personal.

So, the dream state reality is consciousness-driven or consciousness-generated, as opposed to being an objective reality that exists on its own and is only experienced by your consciousness.

On the other hand, we all typically feel that the so-called real world is objective—that it exists on its own and that we "play" in it. However, as we shall see, the most advanced reality experiments in quantum physics have proven beyond reasonable doubt that this reality, which we take for granted, and which feels so fixed and objective, is actually not so at all. It turns out that we can consciously modify this reality.

What is Real?

All of this calls into question what the word "real" really means. Since science has shown that objective reality does not exist (see the "Consensus Reality" section below), the words "real" and "reality" really (pardon the pun) need to be redefined. Please recall the definitions given in Chapter 3. "Real" in this new context means "that which we actually experience"—all experiences, including dreams, OBEs and our waking reality. So the *experiences* are real, but the nature of our reality? Perhaps not so much.

Consensus Reality

The level of consensus is a very important property of a reality—

in fact, a defining property, as we shall see. For example, the only significant difference between a dream state and waking reality is the level of consensus that is applied to it.

When we dream or fantasize, our minds are fully in control of creating the reality that we take part in. In our waking world, however, this is clearly not the case. We can't just fly, make the sky red or defy the laws of physics. However, there is incontrovertible evidence that we can mold our reality, as demonstrated by:

- *The power of the placebo.* For example, placebos were shown to be effective as active treatment in patients with mild neurological deficits, producing an improvement in symptoms of about 50 percent, according to a study by the Bayer Pharmaceutical Research Center.[1]
- *The power of positive intent.* For example, positive emotions have been shown to increase openness to new experiences, according to a study done by the Journal of Consumer Research.[2]
- *The observer effect.* For example, researchers at the Weizmann Institute of Science conducted a highly controlled experiment demonstrating how a beam of electrons is affected by the act of being observed.[3]

And, as if to put the final nail in the materialistic determinism coffin, scientists at the prestigious institute in Vienna developed experiments that demonstrated to a certainty of 1 part in 10^{80} that objective reality does not exist.[4]

So why does waking reality seem so real? It is because it is designed that way. We are much more likely to learn when we believe in well-grounded cause and effect. Seriously, when was the last time you actually consciously learned something from a dream? (Subconsciously, that may be a different story.) In order for us to get something useful out of this physical-matter-reality learning lab, we must believe it is somehow more real than what we can conjure up in our minds. But, again, all that means is that our experience is relatively consistent with that of our free-willed friends and colleagues. She sees a blue car, you see a blue car, you both describe it the same way, and it therefore seems real and objective. Others have referred to this as a consensus reality, a descriptor that fits well.

It is not unlike a large-scale computer game. In a FPS (first person shooter), only you are experiencing the sim. In a MMORPG (massively multiplayer online role playing game), everyone experiences the same sim. However, if you think about it, there is no reason why the game can't present different aspects of the sim to different players based on their attributes or skills. In fact, this is exactly what some games do.

So, one can imagine a spectrum of "consensus influence," with various realities placed somewhere on that spectrum. At the far left, is solipsism—realities that belong to a singular conscious entity. We may give this a consensus factor of 0, since there is none. At the other end of the spectrum is our waking reality, what most of us call "the real world." We can't give it a consensus factor of 100, because of the observer effect. A full consensus factor of 100 would have to be reserved for the concept of a fully deterministic reality—a concept that, like the concept of infinity, only exists in theory. So our waking reality is 99.99-something. Everything else falls in between.

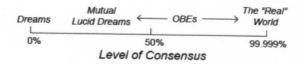

Figure 6.1 The Consensus Reality Spectrum

Many researchers have experienced realities at various points on this spectrum. Individual OBEs that have closely locked into PR are at the high-consensus end of the scale. OBEs that are more fluid are somewhere in the middle. Mutual lucid dreaming can be considered a consensus of two and is therefore somewhere toward the low-consensus side of the spectrum.

Whither Solipsism?

"For the scientist who has lived by his faith in the power of reason, the story ends like a bad dream. He has scaled the mountains of ignorance; he is about to conquer the highest peak; as he pulls himself over the final rock, he is greeted by a band of theologians who have been sitting there for centuries."
- *Robert Jastrow, physicist*

Merriam-Webster defines solipsism as "a theory in philosophy that your own existence is the only thing that is real or that can be known." Philosophically, it is true that your own existence is the only thing that can be known, regardless of the true nature of reality. We can't know for sure whether the other conscious entities in our experience are free-willed ICs like ourselves or elaborate simulations, like robotic zombies. However, a common metaphysical interpretation of solipsism is the idea that the experiencer is the only one who is real in the objective sense of the word. From the evidence presented in this section, free-willed consciousness is responsible for the observer effect and the fundamental creation of reality. What would be the purpose to simulate this? If I were truly living in a solipsistic simulation, wouldn't it be simpler just to make reality look deterministic, rather than driven by the consciousness of others? Of course, this isn't a foolproof argument but a strong one. Plus, if I am the only one with free will, this book is rather pointless, so I have to assume a consensus reality if only to prevent a serious headache.

Therefore, as is clear from the diagram above, purely

subjective dreams are the closest experience that we have to solipsism. But our daily "reality" experience, even if it is virtual, is at the opposite end of the consensus reality spectrum. So it is a mistake to think that simulation theory is solipsistic.

Evidence: Localism, Realism, Free Will and the Observer Effect

"If a man speaks his mind in a forest and no woman hears him, is he still wrong?"
- *Sir Ken Robinson*

The importance of the observer effect and quantum realism cannot be understated and yet, there is so much misinformation out there about it. It is worth taking a walk down history lane to underscore how we got to this point. This section of the book is probably the most important to appreciate, as it is one of the "smoking guns" of Digital Consciousness theory.

Back in the day, there were two theories of light. The popular favorite was Isaac Newton's corpuscular theory of light, which asserted that light was composed of particles or "corpuscles." The competitive theory was Christiaan Huygens' wave theory of light. In 1801, scientist Thomas Young sought to determine once and for all which theory was true, and he developed an experiment now known as the double-slit experiment, which may very well go down in history as the experiment that changed the course of humanity.

In Thomas Young's experiment, sunlight was passed through a screen with a narrow slit (S1 in the figure), so that the resulting light was more coherent (for example, tended to have a narrower range of frequencies). That somewhat coherent light then passed through another screen, S2, this time containing two slits. If the light was wave-based, Young reasoned, the peaks and valleys of the waveforms of the resulting two streams of light (one

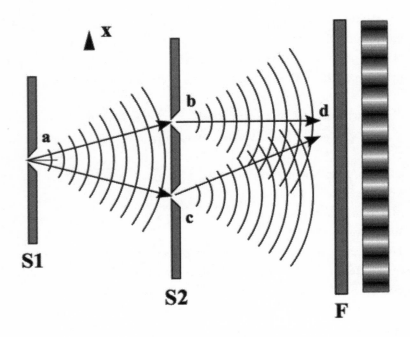

Figure 6.2 Double-slit Experiment[5]

from each slit) should interfere with each other and produce an interference pattern on the final phosphorescent screen (F). This is exactly what was observed and so the corpuscular theory of light fell out of favor.

However, in the early 1900s, with Einstein's work on the photoelectric effect (for which he received the Nobel Prize in physics in 1921), it was shown that light does sometimes behave like particles in that it interacts with matter in discrete quanta of energy, which were later dubbed "photons." This became known as the "wave-particle duality" nature of light. Sometimes light behaves like waves and sometimes it behaves like particles; or, rather, it has both properties all the time, but which "face" of light is observed depends on the nature of the experiment or measuring equipment. As confirmation, later refinements to the double-slit experiment demonstrated that with a single slit open, the photons behaved like particles and created a single band of marks on the screen.

Further advances in technology allowed for photons to be generated one at a time. The result was the same—an interference pattern, which showed conclusively that the interference was not owing to one photon interfering with another, but rather to the photon interfering with itself. It seemed to imply that each photon knew that both slits were open and knew where to go to contribute to the interference pattern.

The reader may recognize from the quantum mechanics primer in Chapter 5 that this is an example of the quantum superposition principle, whereby a particle can be in multiple states, or positions, at the same time. According to this principle, the particle goes through both slits at the same time. Another way to look at it is that at the point in time where the particle goes through the first screen, it is in two positions at the same time. According to quantum theory, once it has been determined which slit the particle actually went through (via a measurement), the superposition state (also known as the "probability wave function") collapses. The effect of that collapse is that the particle (or, rather, the part of the particle wave function) that went through the detected slit is really there at that point in time, while the particle (part of the particle wave function) that went through the undetected slit is not really there at the same point in time.

The important thing to remember here is that, as we discussed in Chapter 3, concepts like superposition and probability wave functions are just models that describe behavior; they aren't laws or facts or even necessarily an accurate representation of reality.

In order to verify theory with an experiment, physicists then put a detector before each slit, the point of which was to measure which slit each particle went through. If the experimental results matched quantum theory, the detection of the position of the particle would "collapse" the wave function and the interference pattern would disappear. This is exactly what was observed. In other words, the act of measuring or observing which slit the particle went through causes the outcome of the experiment to change. This is the so-called observer effect and the implication is huge: that *matter doesn't exist in a material state until observed.* The age-old philosophical debate about materialism versus idealism seemed to be tipping in favor of idealism, in that a conscious observer appeared to be an integral part of the creation of experimental results or "reality."

However, most physicists were reluctant to let go of their ingrained materialism mentality. They argued that the detector must somehow be influencing the outcome of the experiment, like taking away energy from the photons, and therefore causing the wavelike outcome to change. However, over the years, less and less intrusive detector methods were applied to the experiment, all with the same results.

Physicist John Wheeler developed an extreme thought experiment in the late 1970s that went like this: What if the detection of the slit that the particles go through is actually made after they create the interference pattern on the screen? According to the collapsing wave function theory (also known as the Copenhagen interpretation of quantum mechanics), even measuring the particle path *after* the wave pattern is created on the screen should cause the wave function to collapse and register two bands of marks on the screen. If this were true, quantum mechanics would appear to violate basic causality, by generating an effect before the cause. Amazingly, in 2000, an experiment was done by Yoon-Ho Kim, R. Yu, S.P. Kulik, Y.H. Shih and Marlan O. Scully, which demonstrated this exact effect and eliminated forever the argument that a detector could be influencing the outcome of the experiment.[6] More astonishingly, quantum behavior now seemed to include a potential element of retrocausality or apparent "pre-cognition" on the part of the particles themselves.

Einstein was perhaps the most well-known challenger to the traditional (Copenhagen) interpretation of quantum mechanics. He was bothered by the whole idea of randomness and the lack of definitive properties until measurement, asserting that there must be deterministic "hidden variables" that accounted for this strange behavior. In 1935, along with Boris Podolsky and Nathan Rosen, he developed a thought experiment around the idea of taking two particles in superposition and separating them arbitrarily far apart from each other, now known as "entanglement." The change of state of both particles simultaneously owing to a measurement on one of them belies the foundations of Einstein's baby, relativity, in that communication between any two distant points can't occur faster than the speed of light; in other words, simultaneity is impossible. Einstein famously referred to this as "spooky action at a distance," poking fun at the idea and implying that there must be hidden variables that would explain this behavior.

[Note: As it turns out, I believe that Einstein was both right and wrong in a sense. The "hidden variables" that he referred to may be the finite state machine of the data that is the particles. (See chapter 8 for the definition of a finite state machine.) Yet, relativity may be ultimately shown to be inaccurate in that information in the form of state changes can travel faster than light. In fact, it is a prediction of Digital Consciousness that FTL (faster than light) interaction will be demonstrated in any way that can be implemented by the change of data structures in the same clock cycle.]

In 1929, French physicist Louis-Victor de Broglie received the Nobel Prize in physics for his hypothesis that the wave-particle duality applies to all matter, not just light. Electron diffraction experiments confirmed his hypothesis by showing that electrons do behave like waves in certain circumstances. But it wasn't until 1961 that the double-slit experiment was actually performed with electrons instead of light. The same results were observed as with light; that is, an interference pattern developed on the screen behind the two slits—further confirmation of the wave-particle duality of matter. Again, by closing one of the two slits, the electrons would act as particles and create a single band of impacts on the detector screen. Furthermore, in 1974, Italian physicists Pier Giorgio Merli, Gian Franco Missiroli and Giulio Pozzi performed the double-slit experiment by firing individual electrons one at a time toward the slits and, over time, allowing the resulting pattern to build up on the screen. This ensured that the interference pattern formed could not be the result of the accumulation of interferences between individual electron pairs. This experiment was voted "the most beautiful experiment in physics" by readers of *Physics World* in 2002.[7]

Hence, these mysterious "non-classical" quantum effects are not limited to photons, but a large class of particles, including, eventually, molecules, such as Buckyballs (a 60-atom molecular configuration of carbon).

Note: As another prediction of Digital Consciousness theory, there is no macroscopic limit to the size of the object that will generate quantum effects. It is just that the larger, more massive the object is, the more sensitive the measurement equipment would need to be to see its wavelike interference pattern. Theoretically, a piano generates a wave function.

But, back to our story...

First, because we are now at a point where the story starts to get extremely controversial (and mind-blowing), a few more definitions are critical to understanding the extent of quantum mechanics:

- Locality (or local causality)—This is the idea that things can only affect each other and impart forces or influences on each other according to classical laws of mechanics. If you punch someone in the face, that is a local interaction. But if you punch the air and imagine that you are impacting someone extremely irritating (for example, Justin Bieber or Donald Trump), and 1,000 miles away Bieber or Trump suddenly feels the full force of your fist, that would be a non-local interaction. These forces or influences can't propagate faster than relativity or they would be considered (relativistically) non-local.

- Realism—This has been defined before, but it has a central place in the theories of quantum mechanics and the observer effect. It simply means that there is an objective reality, which exists prior to and independently of observations.
- Free Will—The ability to make a choice about something based on more than the history of the universe up to that point.

In 1964, physicist John Bell developed a theorem that states, "No local realistic explanation of quantum mechanical predictions is possible in which the experimenter has a freedom to choose between different measurement settings." Essentially, he argued that certain quantum correlations, like entanglement, necessarily violate realism, locality or free will. This ruled out the possibility of local hidden variables of the type that Einstein envisioned. Still, it was only a theory, and was left to experimental physics and the advancement of technology to be proven.[8]

What does all of this have to do with our theory? Simply this—if it can be shown that local realism is false and free will exists, then a consciousness separate from the body has to exist. I have already argued in *The Universe—Solved!* that free will almost certainly exists. (In short, free will is simply a fundamental aspect of consciousness. Consciousness makes no sense without it.) Therefore, we need to explore the evidence against local realism. Here, the evidence is incredibly strong and getting stronger with every new experiment that probes reality. A brief history...

As it turned out, thoroughly convincing "Bell test" experiments have been very challenging. The general format involves the creation of pairs of photons (entangled because they are created together) that are sent in opposite directions, polarizers that only pass certain orientations (the property that is entangled) and detectors. Initial results demonstrated the quantum theory predictions and seemed to preclude the possibility of local hidden variables (thereby providing supporting evidence against local realism). However, all sorts of loopholes were identified, with esoteric designations such as "detection" (a high enough percentage of samples must be detected to be statistically significant), "locality" (the experimental setup must exclude the possibility of the two measured photons being able to send sub-light speed communication to each other in the time it took to detect their partner's orientation) and "memory" (correlations between sequential samples). One by one, these loopholes have been closed over the years...

In 1982, French physicist Alain Aspect carried out a series of experiments that supported non-local realism, thereby closing the communication (locality) loophole and providing strong evidence against Einstein's hidden variable theory.

In 2003, Nobel Prize-winning physicist Anthony James Leggett developed another set of inequalities, similar to Bell's but based on the idea of non-local hidden variables as a possible solution to the quantum mechanical weirdness.

But then, in 2007, physicist Anton Zeilinger and his colleagues at the Austrian physics institute IQOQI (Institute for Quantum Optics and Quantum Information) performed a set of experiments that violated Leggett's inequality, thereby closing the door on the non-local hidden variable theory. Essentially, what this amounted to was proof that objective reality doesn't exist. The error bar of uncertainty in these results was 1 in 10^{80}.[9]

Over the years, the double-slit experiment has been progressively refined to the point where most of the materialistic arguments have been eliminated. For example, there is now the delayed choice quantum eraser experiment, which puts the "which way" detectors after the interference screen, making it impossible for the detector to interfere physically with the outcome of the experiment. And, one by one, all of the hidden variable possibilities and loopholes have been disproven.

In 2013, a number of experiments definitively closed the detection loophole. And, finally, in 2015, several experiments were performed independently that closed all three loopholes simultaneously with both photons and electrons.[10, 11, 12] Since all of these various experimental tests over the years have shown that realism is false and non-local given the experimenters' choices, the only other explanation could be what John Bell called superdeterminism—a universe completely devoid of free will, running like clockwork while playing out a fully predetermined script of events. If true, this would bring about the extremely odd result that the universe is set up to ensure that the outcomes of these experiments imply the opposite to how the universe really works.

"We always implicitly assume the freedom of the experimentalist... This fundamental assumption is essential to doing science.

If this were not true, then, I suggest, it would make no sense at all to ask Nature questions in an experiment, since then Nature could determine what our questions are, and that could guide our questions such that we arrive at a false picture of Nature."
- *Physicist Anton Zeilinger*

No scientist wants to believe in superdeterminism and most do not. The very idea not only defeats the purpose of science, but it effectively makes life (and reading this book) completely pointless. As free will is core to the nature of consciousness, all of the other categories of evidence in this chapter support its existence.

The net result is that materialism-based theories on reality are being chipped away experiment by experiment. Those who believe in materialist dogma are finding themselves being painted into an ever-shrinking philosophical corner. But idealism-based theories are huge with possibilities, very few of which have been falsified experimentally.

Tom Campbell has boldly suggested a number of double-slit experiments that can probe the nature of reality a little deeper. Campbell, like me, believes that consciousness plays a key role in the nature and creation of our reality. So much so that he believes that the outcome of the double-slit experiments is strictly owing to the conscious observation of the which-way detector data. In other words, if no human (or "sufficiently conscious" entity) observes the data, the interference pattern should remain. Theoretically, one could save the data to a file, store the file on a disk, hide the disk in a box and the interference pattern would remain on the screen. Open the box a day later and the interference pattern should automatically disappear, effectively rewriting history with the knowledge of the paths of the particles.

His ideas have incurred the wrath of the physics trolls, who are quick to point out that regardless of the fact that humans ever read the data, the interference pattern is gone if the detectors record the data. The data can be destroyed, or not even written to a permanent medium, and the interference pattern would be gone. If these claims are true, it does not prove materialism at all. But it does infer something very interesting.

From this and many other categories of evidence, it is strongly likely that our reality is being generated dynamically. Quantum entanglement, quantum Zeno effect and the observer effect all look very much like artifacts of an efficient system that dynamically creates reality *as needed*. It is the *"as needed"* part of this assertion that is most interesting.

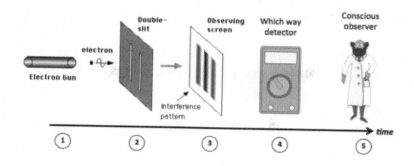

Figure 6.3

If the physicists are right, reality comes into existence at point 4 in Figure 6.3. Why would that be? The paths of the particles are apparently not *needed* for the experience of the conscious observer, but rather to satisfy the consistency of the experiment. The fact that the detector registers the data is enough to create the reality. Perhaps the system "realizes" that it is less efficient to leave hanging experiments all over the place until a human "opens the envelope" than it is to instantiate real electron paths despite the unlikely possibility of data deletion. But it also indicates a sophisticated awareness of all of the probabilities of how the reality can play out vis-à-vis potential human interactions. As such, going forward in this book, the notion of "observation" (for example, "observer effect") shall include any need to satisfy a measurement.

The net result is that the observer effect and the strong Bell test experimental results imply that consciousness is not just fundamental to our subjective reality, but that it is also a creative force behind it.

Evidence: Evolution

When he was a teenager, cognitive scientist Donald Hoffman, like many of us, pondered the idea of humans as machines. He was faced with the stark contrast of the scientific point of view which, as he learned in school, said that humans are essentially machines, and that of the church where his father was a minister, which taught that we are not. His quest for answers led him to research at the Laboratory for Artificial Intelligence at the Massachusetts Institute of Technology, and ultimately a career in Cognitive Science and Consciousness. He has developed a theory of consciousness that asserts that evolution has led us to be blind to a true reality. As he says, "Evolution has shaped us with perceptions that allow us to survive. But part of that involves hiding from us the stuff we don't need to know. And that's pretty much all of reality, whatever reality might be. If you had to spend all that time figuring it out, the tiger would eat you."[13] Essentially, while there is an evolutionary advantage toward tuning our perceptions to anything that supports "fitness" or survival, the actual real-world structures don't necessarily match the associated fitness function.

This theory has support from evolutionary simulations as well as mathematical proofs of some of the concepts. Hoffman developed a mathematical consciousness model, which consists of "spaces" of artifacts, experiences and mappings:

> I have a space X of experiences, a space G of actions, and an algorithm D that lets me choose a new action given my experiences. Then I posited a W for a world, which is also a probability space. Somehow the world affects my perceptions, so there's a perception map P from the world to my experiences, and when I act, I change the world, so there's a map A from the space of actions to the world.[13]

Most interestingly, the math and supporting evidence for this model leads to the conclusion that multiple conscious agents can combine to form a unified consciousness and that W, the true reality, appears to act exactly as a unified consciousness.

This sounds uncannily similar to the idea of ATTI or the global consciousness system which, in this model, would exactly be W.

Evidence: Near-death and Shared-death Experiences

"[Neuroscientists] are mystified as to why they don't make progress. They don't avail themselves of the incredible insights and breakthroughs that physics has made. Those insights are out there for us to use, and yet my field says, 'We'll stick with Newton, thank you. We'll stay 300 years behind in our physics.'"
- *Cognitive Scientist Donald Hoffman*

A near-death experience or NDE is a special state of consciousness that most frequently occurs during near-death circumstances. It is typically characterized by the inclusion of a number of the following elements:

- Out-of-body experience; viewing your body from above
- Traveling at high speed through a tunnel toward a light
- Entering a region of bright light and warmth
- Greeted by deceased relatives and/or friends
- Overwhelming feeling of joy and peace
- Meeting of spiritual guide, or "light beings"
- A life review (quick review of all life events)
- Either making a decision to go back (to Earth) or being told your time is not yet "up"

According to a 1991 Gallup poll, more than 13 million Americans have had NDEs. Based on various studies, people of all religious beliefs or non-beliefs are just as likely to have one. NDEs have been recorded as far back as Ancient Greece, Egypt and India, and appear to be very similar to those recorded today.[14]

It is most interesting that no one who experiences an NDE

ever meets and communicates with people who are alive. No materialistic explanation for NDEs can explain this anomaly and it is clearly statistically significant. In one case, the face of the man seen was discovered 10 years later to be the biological father of the experiencer. Even children who have NDEs see deceased grandparents rather than their own parents, which clearly refutes the skeptics' claims that such experiences might only be wishful thinking.

Some NDEs actually occur during an acute fear-of-death situation rather than during a near-death situation. The fact that a very lucid consciousness occurs when all brain function has ceased challenges the materialist view that consciousness arises from the brain. And the most commonly claimed explanation by skeptics, oxygen deficiency, is completely ruled out as a cause of NDE for those cases of fear-death experience.

Furthermore, many people have experienced what are known as shared-death experiences (SDEs), which have all of the characteristics of NDEs, except that they happen to someone who is present with a person who is dying. Such experiences have been researched since the late 1800s. In researching end-of-life phenomena, neurophysiologist Peter Fenwick and his wife, Elizabeth, collected hundreds of such experiences in the UK and northern Europe. Again, oxygen deprivation can be ruled out completely as an explanation in these cases.[15]

In 2014, the results of a 4-year study in "awareness during resuscitation" was released by scientists at the University of Southampton in the UK. The study was based on the experiences of over 2,000 cardiac arrest patients from 15 different hospitals in the UK, the US and Austria, making it the largest study of its kind ever conducted. Of all of the surviving patients, 39 percent experienced some form of "awareness" even during the time that they were clinically dead; that is, with no heartbeat and no registered brain activity.

One patient witnessed his resuscitation from an "out-of-body" (OBE) state in the upper corner of the room 3 minutes after being clinically dead. Regarding that case, the lead author of the study, Sam Parnia, Assistant Professor of Critical Care Medicine and Director of Resuscitation Research at the State University of New York at Stony Brook, USA, reported to *The Telegraph* newspaper:

> We know the brain can't function when the heart has stopped beating... But in this case, conscious awareness appears to have continued for up to three minutes into the period when the heart wasn't beating, even though the brain typically shuts down within 20-30 seconds after the heart has stopped... The man described everything that had happened in the room, but importantly, he heard two bleeps from a machine that makes a noise at three-minute intervals. So we could time how long the experienced lasted for... He seemed very credible and everything that he said had happened to him had actually happened.[16]

The specific circumstances of this case also refute another common claim among skeptics—that the NDE experience itself occurs after the subject regains brain function.

It turns out that it is fairly common to have otherworldly experiences under extreme near-death conditions. One oft-quoted case involves the singer Pam Reynolds, who suffered a brain aneurysm in 1991 and underwent an operation to have it removed. Neurosurgeon Robert Spetzler performed what is known as a "deep hypothermic circulatory arrest" procedure on her, whereby she was anesthetized, her eyes taped shut and a device generating a 100-decibel "click" placed in her ear in order to measure brain stem activity. At the point when they cut into her skull, she went into an OBE state, recalling conversations and seeing things that should have been impossible for her experience. Next, her blood was drained from her body, her EEG flatlined, her body cooled to 60 degrees and she was clinically dead. Yet, at that point, she had a full NDE that, coupled with the initial OBE, was a continuous experience. The surgeon stated that he had no idea how it could have happened.[17]

"I think death is an illusion. I think death is a really nasty bad lie."
- *NDE experiencer Pam Reynolds*

There were 42 studies of NDEs between 1975 and 2005, covering over 2,500 patients. A meta-analysis of these studies reveals similar results as the Southampton study mentioned above. Most of these studies were "retrospective," in that the respondents to the study self-selected. This doesn't mean that any of the outcomes are necessarily in question, but a more scientific approach to this kind of research is the use of "prospective" studies, which don't involve self-selection but include a control group and independent verifications. In prospective studies, 11–18 percent of cardiac arrest survivors consistently reported NDEs.

NDEs have been prompted by comas, cardiac arrest, anesthesia and unconsciousness for various reasons, which cause serious brain function impairment. However, they have also been triggered by non-impaired brain function events like isolation, depression, existential crisis, meditation, fear-of-death situations or walks in nature.[18] In 1988, Dutch cardiologist Pim van Lommel initiated the first large-scale prospective NDE study in the world. Follow-up studies 2 and 8 years after the cardiac arrest were conducted on survivors who had NDEs, as well as on a control group of cardiac arrest survivors who did not. There were 344 patients who had 509 resuscitations included in the study, which was published in the prestigious medical journal *The Lancet* in 2001. It tracked the frequency of various NDE elements in 62 patients, as well as all of the medical circumstances surrounding each incident. As a result, it was determined that medical causes could not be responsible for the experiences or everyone would have had them. Neither could psychological causes be to blame, because there was no difference between the likelihood of having an NDE or not having one based on a prior fear of death. And since the medication administered had no effect on the likelihood of having an NDE, pharmacological causes could also be ruled out as causation for the experiences.

In fact, the results of the study showed that the individuals who had the NDEs experienced significant life changes in the years following the experience as compared to those who suffered cardiac arrests and did not have an NDE. For example, statistically there was a significant increase in the likelihood of all of the following changes after two years: understanding oneself, appreciating ordinary things, decreased fear of death, showing emotions, accepting others, being more loving and empathetic, understanding others, sensing the inner meaning of life, believing in life after death, having interest in the meaning of life and spirituality, understanding the purpose of life, and an increased involvement with family.

It is hard to explain these changes as being the result of the brush with death itself, as they only correlate strongly with the NDE experience. The people who had the NDEs also generally reported a heightened increase in intuition, connectedness with nature and paranormal gifts. In the US and the UK, three other similar prospective studies were conducted with virtually identical results, for a total of 562 patients. No traditional scientific explanation can fit the results of any of these studies.[19]

There is no shortage of brave and open-minded medical doctors and research facilities that, after conducting the scientific research, support the idea of an afterlife. Dr. Gary Schwartz has a PhD from Harvard University, Cambridge, MA, and was a professor of psychiatry and psychology at Yale University, New Haven, Connecticut. He is currently a professor of psychology, medicine, neurology, psychiatry and surgery at the University of Arizona and the director of its Laboratory for Advances in Consciousness and Health. His book *The Afterlife Experiments: Breakthrough Scientific Evidence of Life After Death* documents the results of known mediums attempting to contact the dead friends of a number of individuals ("sitters").[20] In some cases, the sitters were completely silent and in other cases, they could speak only "yes" or "no." The mediums' accuracy score for both the HBO Dream-Team experiment and the Miraval Silent-Sitter experiment (in which no words were spoken) was 77 percent as compared with 36 percent for control subjects. The data indicated a probability against chance of less than 1 in 10 million, with the net result being "definite examples of precognition and surprisingly accurate observations by the mediums," according to William Beatty from Booklist Review. Explanations include telepathy and the survival of consciousness after death.

Dr. Eben Alexander III is a neurosurgeon with a distinguished career in many of the finest hospitals in Boston, MA. In 2012, he wrote a book, *Proof of Heaven: A Neurosurgeon's Journey into the Afterlife*, about his own NDE and his conviction in the existence of an afterlife.[21] In his article "My Experience in Coma," written for the *AANS Neurosurgeon* medical journal, he wrote:

> My coma taught me many things. First and foremost, near-death experiences, and related mystical states of awareness, reveal crucial truths about the nature of existence. And the reductive materialist (physicalist) model, on which conventional science is based, is fundamentally flawed. At its core, it intentionally ignores what I believe is the fundament of all existence—the nature of consciousness.[22]

Edward F. Kelly, PhD, is currently a research professor in the Department of Psychiatric Medicine at the University of Virginia. Along with researchers Adam Crabtree PhD (therapeutic counseling), Michael Grosso, PhD (philosophy), Dr. Alan Gauld and Dr. Emily Williams Kelly from the University of Virginia, he authored the book *Irreducible Mind: Toward a Psychology for the 21st Century*, which compiles comprehensive evidence for a variety of psychological phenomena "that are extremely difficult, and in some cases clearly impossible, to account for in conventional physicalist terms."[23] Regarding NDE research, the authors conclude that such "complex states of consciousness, including vivid mentation, sensory perception and memory" are nearly impossible to explain, given that "current neurophysiological models of the production of mind by brain deem such states impossible."[24]

Dr. Kenneth Ring is Professor Emeritus of Psychology at the University of Connecticut, and has written dozens of papers and books on his research into the NDE. His 2008 book, *Mindsight: Near-Death and Out-of-Body Experiences in the Blind*, written with colleague and scientist Sharon Cooper, investigates 31 blind people (14 were blind from birth) who had NDEs. Most of the subjects reported visual experiences unlike any they had ever had before. The experiences were reported to be a richer form of visual perception than our typical three-dimensional perception: more of a "knowing" or an awareness of the visual perception of their environment from all angles.

What is interesting about these results is the contrast to dream experiences of congenitally blind individuals. People tend to dream about the sensations that they are familiar with. For this reason, congenitally blind people do not have visual sensations when they dream.[25] Again, materialist interpretations fail at explaining how visual perception can occur without ever having had visual sensory abilities.[26]

While this section has focused thus far on the classic NDE, there are other common end-of-life phenomena that also defy traditional explanations. Peter Fenwick, MD, and Elizabeth Fenwick, RN, have researched hundreds of SDEs, whereby loved ones or caregivers share in a dying person's experience. Common elements to the SDE include mist at death, hearing beautiful music, a change in the geometry of the room, feeling a strong upward pull on the body, sharing an OBE, seeing a mystical light, empathically co-living the life review of the dying person, being greeted by beings of light, and encountering heavenly realms and a boundary in that realm.[27] People who have such experiences often have the same lasting effects (for example, reduced fear of death and greater understanding of life's purpose) as those who directly experience NDEs.

Another well-studied phenomenon is deathbed communication, which includes the more familiar "deathbed visions," but the distinction is made because some people only have auditory perceptions. According to a study by Dr. Madelaine Lawrence and Dr. Elizabeth Repede, published in the *American Journal of Hospice & Palliative Medicine*, the average hospice nurse sees about five patients per month who experience deathbed communications.[28] The people encountered in these visions are invariably dead and in some cases, they are people who are not even known to be dead by the experiencer. Cases have been reported (and I have personally experienced this) of animals behaving strangely, howling or even focusing at the same point in the room as the dying person, as if they are seeing something that we do not.[29]

In summary, the collective evidence on NDEs (corroborating artifacts, temporal markers, similar experiences under conditions without impaired brain function, life-changing impacts, etc.) rules out every single materialist explanation that the skeptics can come up with. The only conclusion that can be drawn is that at least some aspect of consciousness exists outside of brain function.

Evidence: Out-of-body Experiences

"In 1958, without any apparent cause, I began to float out of my physical body. It was not voluntary; I was not attempting any mental feats. It was not during sleep, so I couldn't dismiss it as simply a dream. I had full, conscious awareness

of what was happening, which of course only made it worse. I assumed it was some sort of hallucination caused by something dangerous—a brain tumor, or impending mental illness. Or imminent death... It occurred usually when I would lie down or relax for rest or preparatory to sleep—not every time but several times weekly. I would float up a few feet above my body before I became aware of what was happening. Terrified, I would struggle through the air and back into my physical body. Try as I might, I could not prevent it from recurring."[30]
- *Robert Monroe*

Robert Monroe, born in Indiana in 1915, had a distinguished career as a writer, director and composer in the radio broadcasting industry, and eventually as a radio broadcasting executive, forming his own production company, and serving on the board of directors for the Mutual Broadcasting System network. His company owned a number of radio stations and later branched into cable television.

In 1956, as a result of his interest in human learning and consciousness, he established a research and development division of his company in order to study the possibilities of learning while sleeping. He volunteered to be the test subject for some of the experiments and in 1958, he began to experience a state of consciousness that was separate from his body.

This experience happened a number of times and ultimately culminated into what he referred to as an "out-of-body experience" (OOBE or OBE). This terminology stuck and while not unknown at the time, it provided experimental validation for experiences that many other people had had. Monroe himself observed, in his book *Far Journeys*, that around 25 percent of people surveyed remember having such an experience at some point in their lives.

In the early 1970s, after publishing his first book *Journeys Out of the Body*, Monroe recruited a number of scientists and engineers to help him investigate and formalize methods for repeatably inducing OBEs. Physicist Tom Campbell, who had worked at NASA and the Department of Defense, joined Monroe's team at Monroe Laboratories at this time and embarked on his journey of seeking the truth about the nature of reality. As Campbell recollects in his book *My Big TOE*, they were like "mad scientists" trying all sorts of techniques to stimulate the experience.

Based on some research that his partner, electrical engineer Dennis Mennerich, did, they hit upon the idea of using two different tones, one injected in each ear, that differ in frequency by 4 Hz. They chose this frequency owing to Monroe's observation of feeling a vibration at that rate just before going out of body. The results were very successful and resulted in fairly consistent OBEs. In order to determine if the OBEs were actual experiences in a "real" place, as opposed to being simply fantasies of the mind, Monroe's team designed scientifically rigorous experiments that eliminated the fantasy possibilities. For example, having two subjects in separate soundproof rooms travel out of body at the same time, each with the intent to find each other in the "non-physical" realm, and then communicate with each other and travel together, established that the experiences were not purely a single person's fantasy. Each subject took careful notes after the experience and before leaving the soundproof room, after which notes were compared.

These experiments were statistically significant in their conclusion that multiple people were sharing a non-physical reality experience. However, it still left the possibility that one person had the fantasy and the other perceived it or shared in it through telepathy. Although this possibility would be in and of itself astounding, it was also ruled out by having a third party write things on a chalkboard while the other two were in the altered state, and then erase what they wrote. The subjects correctly observed such corroborating evidence of true OBEs.

These experiences were not the first OBEs reported, they were just the more rigorously analyzed and they pushed the boundaries of repeatability, and the demonstration of additional reality levels. Other OBE studies such as Dr. Charles Tart's (University of California, Davis) "Miss Z" studies also demonstrated true OBEs via the use of corroborating evidence.

While millions of experiencers continue to travel out of body, science is slowly catching up but methodically resisting the implications. A study was published in 2014 in *Frontiers in Neuroscience* by the School of Psychology at the University of Ottawa, focusing on the analysis of fMRI brain scans during OBEs that a subject was apparently able to induce at will.[31] For some reason, the researchers felt the need to change the name of the experience from the conventional OBE to "ECE" (extra-corporeal experience) and the report is full of references to "illusions," which of course represents materialistic scientific bias. Layperson websites and magazines that picked up the story did no better as, for example, Gizmodo reported, "It *is* real in the sense that she's actually experiencing it. The brain scans show that she's going through what she's claiming. But that doesn't mean that her 'soul' is getting out of her body. This is *not* an astral trip, like those described by mystics. There's *no* paranormal activity of any kind."[32]

Really? On what basis does this magazine make these definitive claims? It's the old "there's electrical activity in the TV so the programming must come from inside the television" argument. I love how they emphasize "not" and "no" with absolutely *no* clue what they are talking about. Some things will never change.

Evidence: Past-life Experiences

"I have been born more times than anybody except Krishna."
- *Mark Twain*

There is nothing new about reincarnation. It is central to the world's most ancient religions and belief systems.

> "Just as a man discards worn out clothes and puts on new clothes, the soul discards worn out bodies and wears new ones."
> - *Bhagavad Gita (2.22)*

The verse above comes from the Bhagavad Gita, a Hindu religious text purportedly spoken by the god Krishna to his friend Arjuna some 5,000 years ago. These texts were written like poetry and were passed down orally for thousands of years before being codified in writing around 500 BCE. According to the text, Krishna taught the Bhagavad Gita to the sun god Vivasvan 120 million years ago and it has existed in human society for 2 million years.[33] Regardless of whether it is 120 million, 2 million, 5,000 or 2,500 years old, reincarnation is one of the most ancient philosophical concepts in human society. It is also ubiquitous throughout the world.

Western Europe—"the Pythagorean doctrine prevails among the Gauls' teaching that the souls of men are immortal, and that after a fixed number of years they will enter into another body."[34] Greek scholar Cornelius Alexander Polyhistor (1st century BCE)

British Isles—"the principal point of their [Celts, Druids] doctrine is that the soul does not die and that after death it passes from one body into another."[35] Julius Caesar (1st century BCE)

Ancient Egypt—"The Egyptians were the first who asserted that the soul of man is immortal, and that when the body perishes it enters into some other animal, constantly springing into existence."[36] Herodotus (5th century BCE)

Ancient China—Possibly as far back as 2500 BCE, the I Ching referred to "eternal cycles of life and multidimensional development that could involve many lifetimes." Taoist philosophy concludes: "Birth is not a beginning; Death is not an end. There is existence without limitation; there is continuity without a starting-point."[37] Chuang Tzu, Chinese philosopher (4th century BCE)

Africa—According to researcher Paul Von Ward, the Yoruba people of West Africa believed that reincarnation often happened within the family. "A child born in a family where the death of a grandparent had recently preceded its birth was sometimes thought of as the re-birth of that being. The child might be called Babatunde for 'Father has returned' or Yetunde for 'Mother has returned.'"[38] Further, researcher Andrew Rooke notes that belief in rebirth is part of many cultures throughout the entire continent, including Akamba (Kenya), Akan (Ghana), Lango (Uganda), Luo (Zambia), Ndebele (Zimbabwe), Sebei (Uganda), Shona (Zimbabwe), Nupe (Nigeria) and Illa (Zambia).[39]

Buddhist regions—As is well known, reincarnation is a central tenet of Buddhist philosophy, which teaches that reincarnation cycles called saṃsāra are essentially endless. The region covered includes almost all of the Far East.

Hinduism, Jainism and Sikhism—These are all religions or philosophies that are based in part on reincarnation, and cover India, mid-Asia and parts of the Near East. Combined, these consist of over 1 billion adherents.

Siberian, Mongolian and Native American shamanism—Mongolian and Siberian shamans believe in and teach the reincarnation of human souls. These groups represent millions of people across huge geographical regions.[40] It is also well established that many Native American cultures believe in reincarnation. For example, Gitxsan writer Shirley Muldon reports: "We believe in reincarnation of people and animals. We believe that the dead can visit this world and that the living can enter the past. We believe that memory survives from generation to generation. Our elders remember the past because they have lived it."[41] Because of the general lack of ancient written records, it is difficult to identify the origins of the belief systems. However, it is striking that many of the same teachings occur in Siberian, Mongolian, and Native North and South American cultures, especially considering that DNA evidence has clearly demonstrated a link between the cultures that extends prior to the last ice age.

Aboriginal shamanism — According to researchers, despite the separation of Australian native peoples into hundreds of disparate tribes, reincarnation remains a common belief across all of them.[42]

Traditional religions were severely suppressed in many countries during the communist era, most notably in Russia and China. According to the 20 July 1987 issue of *Stuttgarter Zeitung*, for example:

> Vice Governor Pu Quiong *reported that before the rebellion of 1959, which led to the flight of the Dalai Lama, there were 2,700 temples and monasteries with* 114,000 monks and 1,600 "living Buddhas". The "democratic reforms" reduced the monasteries to 550 with 6,900 monks [by] 1966. After the turmoil of the Cultural Revolution from 1966 to 1978 [there] were only eight monasteries with 970 monks left.

What is left, in terms of world religions and spiritual thought? It is mainly only the "Abrahamic" religions (Judaism, Christianity and Islam) that do not have a strong concept of reincarnation. Or do they?

Esoteric Judaism — Kabbalistic thought is an esoteric branch of Judaism that is considered by many scholars to predate traditional Abrahamic creation philosophies. Flavius Josephus, a Jewish historian from the 1st century BCE wrote in his historical work *Jewish Antiquities*:

> The bodies of all men are, indeed, mortal, and are created out of corruptible matter; but the soul is ever immortal, and

is a portion of the divinity that inhabits our bodies... Do not you know, that those who depart out of this life according to the laws of nature... enjoy eternal fame; that their houses and posterity are sure; that their souls are pure and obedient, and obtain a most holy place in heaven, from whence, in the revolution of ages, they are again sent... into bodies.[43]

According to Von Ward:

This ancient Hebrew tradition's concept of reincarnation involved working towards a reconnection with the creative source of light. It is not unlike the Hindu/Buddhist notion of achievement of enlightenment (awakening to the light). The Kabbalist saw achievement of enlightenment, "illumination" or "awakening" as rising to the level of a co-creator of the universe—along with other creator-beings.[44]

Esoteric Christianity—Some of the earliest followers of the teacher Jesus, known in Greek as Gnostics, shared this cosmology.[45] He also identified many clear references to reincarnation beliefs directly from the Bible, such as John 9:1, which reads, "The disciples asked Jesus, 'Teacher, whose sin caused him to be born blind? Was it his own or his parents?'" which implies the concept of karma in addition to reincarnation. In fact, most early Christians included the idea of reincarnation in their belief system; that is, until the Council of Nicea in 325 CE. At that time, Roman emperor Constantinople and the Roman religious leadership decided to pick and choose the version of Christian teachings that suited their objectives at the time. This objective was to control the pagan masses through fear by declaring a brilliant set of ideas:

1. Each person only had a single lifetime
2. A savior died for their sins

3. In order to be saved themselves, they must declare allegiance to the Church and accept their belief system

Gnosticism went underground for 1,000 years, and was eventually rooted out, persecuted and destroyed by the Roman Catholic hierarchy in the Albigensian Crusade in the 13th century.

Esoteric Islam — Meanwhile, a new proselytizing belief system was spreading throughout Europe and the Middle East: Islam. While the prevailing view on reincarnation in Islam is that it is not part of the religion, there are many sects who find reincarnation to be an integral part of the Qur'an, including the traditions of Ikwan al-Safa or Ismailism, Hemarih, Tarih, the modern Druze, Alawi, Ahl-e-Hagh, some Sufi orders, and the followers of Ahmad ibn Khabet and Shia Ghulat.[46] Elements from the Qur'an that support reincarnation include: "And you were dead, and He brought you back to life. And He shall cause you to die, and shall bring you back to life, and in the end shall gather you unto Himself" (2:28) and "They [the unbelievers] will say: 'Our Lord! Twice you have caused us death and twice you have given us life. We now confess our sins. Is there any way out [now]?'" (40:11)

Putting all of this together, reincarnation-based belief systems basically covered the globe in the past. The rise in power of the Western monotheistic religions over the past 2,000 years, combined with the rise of Western science, and its focus on materialism and determinism over the past few hundred years, has dealt a temporary blow to the logic and widespread evidence of reincarnation. However, consciousness-centric science, New Age thought and those of us who are putting pieces of the puzzle together (such as the observer effect, quantum mechanics anomalies and true paranormal research) are recognizing that the Western beliefs in either a single earthly life, plus an afterlife, or no afterlife at all, have become antiquated.

OK, so much about the beliefs. What about the evidence?

James Leininger—James Leininger was born in 1998 to parents Bruce and Andrea. At the age of 2, he began having recurring nightmares where he was a WWII pilot being shot down by the Japanese. "Airplane crash! Plane on fire! Little man can't get out!" When asked who the little man was, he clarified, "Me." As he got older, his parents were puzzled with his obsession with airplanes, and especially with his knowledge of Japanese WWII aircraft and archaic plane terminology (for example, "drop tank").

James was not learning this from either talking to anyone, being read to or from listening to television programs that were on in the house. James knew that that Japanese fighters were called "Zekes" and bombers were called "Bettys." He corrected a narrator on a History Channel program that he was watching with his parents, understanding what Japanese "Tonys" and "Zeros" were. He stated that his name used to be James and that he flew a Corsair that took off from a boat named the *Natoma*. He named his GI Joe action figures "Billy," "Leon" and "Walter"—names that neither his parents nor their friends had ever mentioned. When Bruce asked him why he picked those names, James said, "Because that is who met me when I got to heaven."

Andrea's mother suggested that little James may have been experiencing a past life, but Bruce, being a devout Christian, was extremely reluctant to believe such things. His skepticism, however, eventually subsided on the face of the evidence. Bruce got a book on WWII aircraft carriers in the Pacific and leafed through it with his son, who recognized that it was at the battle of Iwo Jima that his plane was shot down and crashed. Upon further research, he found out that there was a James Huston, who was one of 18 pilots from an aircraft carrier called the *Natoma Bay*, who had died during service. Huston was the only pilot who died during the battle of Iwo Jima. Billie Peeler, Leon Connor and Walter Devlin were the names of three *Natoma Bay* pilots who died before Huston. The little boy had also recalled a friend named Jack Larson, who turned out to be a surviving *Natoma*

Bay pilot, who Bruce later visited at his home in Arkansas. The knowledge that little James had of Huston's family members was uncanny and would be inexplicable without invoking reincarnation.

At one point, James Leininger stated that he selected Bruce and Andrea as parents, while he was in the spirit world. He told his father that he saw his parents when they were vacationing at the "big pink hotel" in Hawaii. In fact, Bruce and Andrea had vacationed at the pink Royal Hawaiian Hotel 5 weeks prior to Andrea becoming pregnant with James. Yet, neither Bruce nor Andrea had ever told James about their Hawaiian vacation.[47]

All it takes is a single irrefutable case like this to provide strong evidence for reincarnation. However, there are many others.

Carl Edon — Another example is Carl Edon was born in 1973 in England and, at the age of 3, began obsessively speaking about his previous life as a Nazi airman. He subsequently had visions of being shot down and claimed to have lost his right leg in the crash. Carl came to an untimely demise at the hands of a murderous colleague at the age of 22. A couple years later, a German airplane was unearthed a few hundred yards from the site of his death. The plane had been piloted by a Nazi airman named Heinrich Richter. Many noticed a strong similarity in appearance between Richter and Edon. In addition, Richter had lost his right leg in the crash, just as Edon had stated.[48]

Ian Stevenson — Dr. Ian Stevenson was a McGill University-educated MD who worked as a psychiatrist and research scientist at the University of Virginia School of Medicine. He was also the founder and director of the University of Virginia's Division of Perceptual Studies, a department that made significant contributions into the rigorous investigating of parapsychological phenomena such as reincarnation, NDEs, OBEs, after-death communications, deathbed visions and altered states of consciousness.

Stevenson's interest in reincarnation was triggered by his

investigation of a case of a Sri Lankan child who had remembered a past life. Upon questioning the child, the child's parents and the child's supposed past-life parents, Stevenson became convinced, despite his scientific training and education, that this was a bona fide case of reincarnation. And so began the most prolific research into reincarnation to date.

Dr. Stevenson authored about 300 papers and wrote 14 books on the topic, including a multi-volume 2,268-page work entitled *Reincarnation and Biology: A Contribution to the Etiology of Birthmarks and Birth Defects*. He documented 200 cases of children who recalled past lives, and had memories and birthmarks corresponding to the lives and wounds of those past-life individuals. In another volume of his research, *Twenty Cases Suggestive of Reincarnation*, Stevenson documented cases of responsive xenoglossy, authentic instances of people who could speak a language that they never learned, but suggestive of a past-life incarnation who spoke that language. While Stevenson's prolific reincarnation research is unparalleled, there have been many other scientists who have also contributed to the field, including his successor, child psychiatrist Jim Tucker, whose 2005 book, *Life Before Life: A Scientific Investigation of Children's Memories of Previous Lives*, contained an additional 40 cases of children's past-life memories.

Past-life regressions—Brian Weiss MD was a graduate of Columbia University and Yale Medical School, and is currently Chairman Emeritus of Psychiatry at the Mount Sinai Medical Center in Miami. He was a traditional psychotherapist in 1980, believing in a conservative approach to treatment and only having respect for ideas proven via the scientific method. But then he met a patient who turned his worldview upside down. He referred to her as Catherine, a woman who came to him to seek treatment for panic attacks and anxiety. After 18 months of conventional therapy, he decided to try hypnosis, which he considered to be an acceptable method to help patients remember repressed memories.

For a few sessions, she recalled some childhood memories. Then, during one session, Weiss asked her to "go back to the time from which your symptoms arise" and she began to report experiences from a "previous life" in Ancient Egypt. He was, of course, extremely skeptical and attributed the recollections to a dreamlike state. But, as the sessions continued, Catherine experienced more past lives and Weiss recognized not only the conviction with which she believed her recollections, but also the therapeutic value that they seemed to have.[49]

For some reason, Catherine was ebullient, happy and making much more progress dealing with her issues in just a few short regression sessions than she did in the 18 months of prior psychotherapy. Dr. Weiss recognized the therapeutic value of past-life regression and, over time, came to believe that it was actually based on reincarnation. He has treated and written about many other patients since Catherine, who have experienced past lives under hypnosis. In some cases, the patients recounted details about places that were completely unknown to them, which were later corroborated.

What Weiss found to be the most fascinating, however, were the accounts of "in between states." According to his patients, after death, our consciousness enters a state in between lives, where we meet with so-called spirit guides or masters, who help us to design and choose our next life, in order to achieve some evolutionary progress of our spirit.

In a nutshell, what Brian Weiss (and other past-life regression therapists, who followed him) learned from his patients is what should now be a fairly common story:

- There is a soul and an afterlife.
- When you die, you meet with "spirit guides" in the "astral plane," who help you design your next life, the objective being to improve the universal qualities of your soul.

- Gender may be swapped from life to life.
- The people who play significant roles in your various lives tend to be your soulmates; those who travel from life to life with you. For example, in this life you may be female and have a husband, who may have been your mother or a teacher in a past life, or your twin brother in a future life.
- There are multiple levels to the astral planes. In these realms, there is no time. All happens at the same time. Future lives and past lives all happen simultaneously in the spirit realm. Time is a physical construct.
- Souls enjoy occupying a body, because only then can they have sensory experiences, which helps them to learn and evolve their consciousness.
- The cycle repeats until you achieve spiritual perfection, at which point you may be a spirit guide or become one with the universal spirit; although, admittedly, this seems to conflict with the idea of timelessness.

These ideas came from a completely orthogonal source from the ancient religious beliefs and shamanistic cultures that we have also been investigating. And yet, they arrived at the same conclusions. And, by the way, so did Edgar Cayce, aka "the sleeping prophet," perhaps the most studied and reputable mystic of the modern age.

Cases like those of James Leininger, Carl Edon, the many children studied by the Division of Perceptual Studies at the University of Virginia, and the many cases of corroborated past-life experiences recounted through hypnotic regression, collectively provide substantial evidence for a reality paradigm whereby the core component of an individual's consciousness resides outside of the body. An alternative viewpoint could be that while the immortal soul lives on in a non-physical reality, it somehow transfers to the brain of each individual incarnated entity until death, at which point it returns to the astral plane.

However, as will be seen, this idea, while still profound, failed to explain all of the other categories of evidence that the Digital Consciousness model does with ease.

It should also be noted that the fact that certain memories can live on in between lives is also great evidence for a digital reality. There are other mechanisms that could theoretically store information, but nothing is as efficient as binary. And why would ATTI evolve inefficiently?

Evidence: Psychic Phenomena

In the late 1960s and early 1970s, at the height of the Cold War, US intelligence uncovered a well-funded and apparently successful psychic research program in the Soviet Union to exploit the power of the mind. Fearing a "psychic warfare gap," the US military/ intelligence community approved funding to carry out psychic research at the prestigious think tank Stanford Research Institute, California (SRI). Former CIA director Robert Gates admitted on the *Nightline* TV show in 1995 that the intelligence community invested about $20 million over a 16-year period from 1972 through 1988. Specifically of interest was the ability of the human mind to sense impressions or information about distant targets through concentration. The term "Remote Viewing" was coined by physicists Russell Targ and Harold Puthoff to describe this effect.

There were many successful viewings from this project. In one case, for example, Targ states:

In 1974, my colleague Hal Puthoff and I carried out a demonstration of psychic abilities for the CIA in which Pat Price, a retired police commissioner, described the inside and outside of a secret Soviet weapons laboratory in the far reaches of Siberia — given only the geographical coordinates of latitude and longitude for a reference (that is, with no on-site cooperation). This trial was such a stunning success that we were forced to undergo a formal Congressional investigation to determine if there had been a breach in National Security. Of course, none was ever found, and we were supported by the government for another fifteen years. As I sat with Price in these experiments at SRI, he made the sketch ..., to illustrate his mental impressions of a giant gantry crane that he psychically "saw" rolling back and forth over a building at the target site![50]

In another instance Ingo Swann, one of the principal participants in the SRI study, suggested an experiment whereby he would remote view the planet Jupiter, his observations to be compared to imagery, to be returned by the upcoming NASA *Pioneer* flyby of Jupiter. Swann did his viewing on 27 April 1973 and shocked other participants in the experiment by observing that there was a ring around the planet. At this point in time, of course, this was completely unexpected, as there was no astronomical evidence of the ring. However, a ring around Jupiter was discovered in 1979 by the *Voyager* probe.

Targ recalls a couple additional seemingly incredible results:

> While remote viewing for the CIA and Army Intelligence, SRI psychics found a downed Russian bomber in Africa, reported on the health of American hostages in Iran, described Soviet weapons factories in Siberia, and forecasted a Chinese atomic-bomb test three days before it occurred.
>
> When San Francisco heiress Patricia Hearst was abducted from her home in Berkeley, a psychic with the SRI team was the first to identify the kidnapper and then accurately describe and locate the kidnap car.[51]

Remote viewing was similar to clairvoyance—the psychic ability to visualize things that are not local. The distinction made between the two abilities is that remote viewing was intended to be a methodical protocol. But that doesn't mean that the results of the protocol are consistent or reproducible; they are not.

The problem with studying paranormal or psychic phenomena in general is that they are typically sudden, out-of-the-blue, experiences. Some people may have only one or two such events in their lifetimes, which obviously makes them very difficult to study. However, there can also be very subtle but more consistent experiences. Fortunately, there is a way to study these types of psychic phenomena and it has to do with the idea of statistical significance. So let's have a brief diversion into the fascinating world of statistics! (I can hear the pages rapidly turning already.)

Let's say we flip a penny. The odds that it will come up heads is 50 percent and the odds that it will come up tails is 50 percent. Pretty simple, right? Let's flip it two times in a row. And to use some symbols for the results, let's call tails a 0 and heads a 1. So the possible combination of two flips is 00, 01, 10 and 11. Therefore, the odds that it comes up heads twice in a row is 25 percent, the odds that it comes up tails twice in a row is 25 percent, and the odds that it will be heads one time and tails another is 50 percent. You can repeat this logic for as many flips as you want. As the numbers get higher, things get a little more interesting. There is actually a formula that will calculate the probabilities. It looks like this:

Probability of having k flips turn up heads out of n tries is:

$$p(k,n) = n!/(k!*(n-k)!*(2^n))$$

In this example, the ! symbol indicates a factorial operation (Google it) and ^ indicates raising to a power. So, for example, with 10 flips, there are a total of 2^n or 1,024 combinations of coin landings, with the distributions looking like the following:

k	Combinations	Probability	Cumulative
0	1	.00097	.00097
1	10	.0098	.0107
2	45	.0439	.0547
3	120	.1172	.1719
4	210	.2051	.3740
5	252	.2451	.6230
6	210	.2051	.8281
7	120	.1172	.9453
8	45	.0439	.9892
9	10	.0098	.9990
10	1	.00098	1

Figure 6.4

If you divide each number of combinations by 1,024, you get the probability that the coin will turn up heads exactly that many times in 10 flips. But if you want to know the odds that a coin flip will land at most k times, you have to use something called the cumulative distribution function, which is the last column in the table. The way to read it is as follows: the odds that a coin flip will turn up heads 6 or fewer times is .8281. You'll see in a minute why all of this is important. It turns out that this is very similar to something called the binomial distribution function. As the number of coin flips gets larger, the two ways of calculating the probabilities become very close to the same. Using that binomial distribution function, we can see the odds on a chart like this:

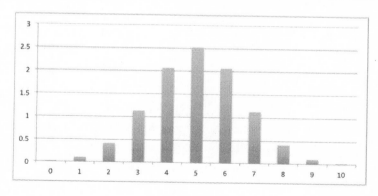

Figure 6.5

As can be seen, the number of times the coin will come up heads 5 times in 10 flips is about 2.5 times. The cumulative distribution function for 10 flips follows:

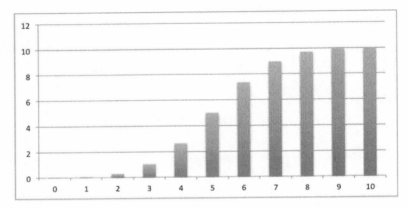

Figure 6.6

Now here is the relevant part. As you take more samples (in other words, flip the coin more times), the odds that it lands about 50 percent of the time increases. So, for example, for 100 flips:

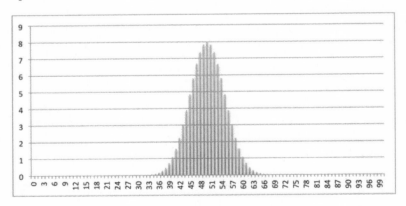

Figure 6.7

And for 1,000 flips:

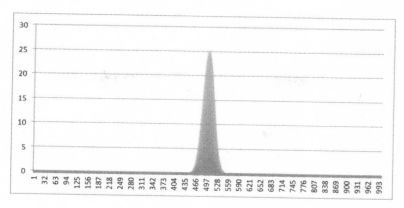

Figure 6.8

Notice how the curve gets narrower as we increase the number of flips. What this means is that the more samples that are taken, the less likely the result will deviate from the average. For example, the odds that a coin flips heads 51 percent or more drops drastically the more times it is attempted:

100 flips—odds of 51% or more heads = 42%
1,000 flips—odds of 51% or more heads = 26%
10,000 flips—odds of 51% or more heads = 2.3%
100,000 flips—odds of 51% or more heads = .000000013%

So, what does all of this have to do with paranormal experiences? Just this—despite the fact that they are not reproducible deterministically, if you take enough samples and the deviation from chance randomness (for example 50 percent on the coin flip) is consistent, then you have found a mathematically and scientifically statistically significant phenomena!

So let's examine the data...

(Note: parts of this section have been reproduced from The Universe—Solved!*)*

Physicist Russell Targ calculated the odds against chance of the SRI remote-viewing experiments yielding the positive results that they did. In the case of the 36 outdoor remote-viewing trials described with 6 US army intelligence officers, their odds against chance of achieving those results was 1 in 3,300.[52]

Over a period of 12 years, Robert G. Jahn, engineering professor and Dean Emeritus at Princeton University, and associate Brenda J. Dunne, conducted over 1,000 experiments, in which subjects concentrated on the outcome of a random event generator and created a statistically significant effect on the results toward the direction of the subjects' intent.[53] While the effects were generally small for each experiment, when taken collectively, the probability of the results occurring naturally were approximately 1 in 3,000,000,000,000. According to Jahn:

> ... it appears that once the illegitimate research and invalid criticism have been set aside, the remaining accumulated evidence of psychic phenomena comprises an array of experimental observations, obtained under reasonable protocols in a variety of scholarly disciplines, which compound to a philosophical dilemma. On one hand, effects inexplicable in terms of established scientific theory, yet having numerous common characteristics, are frequently and widely observed; on the other hand, these effects have so far proven qualitatively and quantitatively irreplaceable, in the strict scientific sense, and appear to be sensitive to a variety of psychological and environmental factors that are difficult to specify, let alone control. Under these circumstances, critical experimentation has been tedious and frustrating at best, and theoretical modeling still searches for vocabulary and concepts, well short of any useful formalisms.[54]

In 1995, Jessica Utts, professor of statistics at the University of California, Davis, completed a report on the analysis of remote viewing at SRI, which shows a statistical significance of 10^{20} to 1 against chance and concludes that remote viewing is a real ability.

In the mid 1990s, at the University of Göteborg in Sweden, the department of psychology conducted experiments in which subjects underwent mild sensory deprivation while "senders" concentrated on randomly chosen videos. In two studies, a very high hit rate of 37 percent was achieved. Higher hit rates were noticed in subjects with a predisposition toward the belief in paranormal effects.[55]

British biologist and author Rupert Sheldrake, in his well-known experiments involving the "sense of being stared at," found that there is a statistical significance to the effect; that is, on average, people can sense when they are being stared at, even when all possible influences are removed. At Our Lady's College in Drogheda, Ireland, for example, in a double-blind experiment involving over 2,000 randomly chosen test points, 57 percent of the subjects were correct in their guess when being stared at. This might not sound like much but, like our penny flipping example, when you have this kind of statistical deviation over so many test cases, it is highly significant. In fact, the chance of this occurring by chance alone is approximately 1 in 1,000,000,000.

These tests were repeated with similar results at the University College School Junior Branch (UCS), a boys' school in Hampstead, London, at Southern Connecticut State University, and at schools in Stuttgart, Hamburg, Bremen, Boca Raton and Stockholm.[56]

Braud and Schlitz conducted 37 studies of direct mental interactions with living systems (DMILS), whereby biological processes were remotely influenced through thought alone. The statistical significance of the results was 1 in 40,000,000,000,000,000.[57]

Dr. Dean Radin, Chief Scientist at the Institute of Noetic Sciences, conducted a comprehensive "meta-analysis" of thousands of independent experiments in telepathy, clairvoyance, perception through time and other psychic phenomena, and compiled all of the results. While any given experiment might not yield remarkable results, when taken as a whole, the body of experimental evidence that supports the existence of a bona fide phenomenon is astounding. Considering telepathy, for example, across 2,549 sessions from 1974 through 1997, the overall "hit rate" had "odds against chance beyond a million billion to one." Very few other scientific fields require that level of certainty before accepting a theory as fact.[58]

And finally, Dr. Daryl Bem, Professor Emeritus of Psychology at Cornell University, conducted a set of precognition studies, the results of which favor precognition with the odds against chance being about 1 in 74 billion.[59]

Oh, and by the way, the odds of being struck by lightning in your lifetime are 1 in 12,000, far less than the odds that these paranormal results are owing to chance. And, more to the point, according to a paper published by Oxford University Press and written by Charles Weiss of Georgetown University, something with a probability of greater than 99 percent (or less than 1 in 100 odds against chance) would be considered "beyond a reasonable doubt" in legal standards, "rigorously proven" by scientific standards and "virtually certain" in the IPCC (Inter-Governmental Panel on Climate Change) scale of scientific certainty.[60] Apparently, paranormal and psychic phenomena have therefore been rigorously proven.

Evidence: Nature vs. Nurture vs. Neither

Nature versus nature—it is an age-old conundrum. Conventional wisdom says that physical traits like height, body shape and eye color are largely the result of genetics, aka "nature." What about intelligence? Susceptibility to diseases? Or personality traits like optimism, responsibility, kindness, laziness, pride or sense of humor. John Locke formulated the idea of tabula rasa, or "blank slate," in the 17th century, which asserted that since we are born with a blank slate of behavioral characteristics, these must be environmental, aka "nurture." Yet, many studies demonstrated that such categorization is rarely that black and white.

For example, psychologists and biologists have attempted to tease out the influential factors by studying criminal records, IQ, personality traits, and sexual preferences of identical twins raised together, identical twins raised apart, adoptive siblings raised together, fraternal twins, siblings and random pairs of strangers. In such a manner, they can identify both the levels of correlation of traits owing to environment and to heredity. For example, see Figure 6.9 below. On the one hand, Trait A shows a high correlation (70 percent) to environment, but a low correlation to heredity, since there is very little difference in the level of correlation between adoptive siblings or identical twins (monozygotic). Trait B, on the other hand, shows a high correlation to heredity but a weak correlation to environment, because the correlation level is strong with twins and weak with adoptive siblings raised in the same environment. Trait C is most interesting, because it seems to be unrelated to environment and heredity.

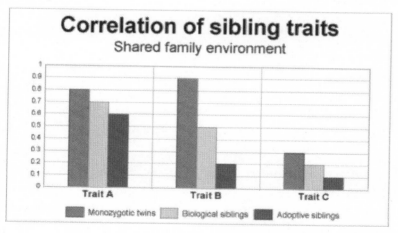

Figure 6.9 (from Wikipedia Commons)[61]

The result of such research indicates that criminality appears to have influences from both nature and nurture, while IQ seems largely hereditary. Most studies support the conclusion that personality traits are mostly hereditary, as adoptive siblings show little similarity in such traits as compared to randomly selected people, whereas identical twins raised separately have a strong correlation on those traits. However, there are many personality traits where there is little genetic correlation and little environmental correlation.

For example, sexual preference appears to be unrelated to DNA, yet is also hard to explain by environment alone, given the results of identical twin studies. However, even in studies of these traits, where correlations are observed, the correlations tend to be small, leaving a large portion of the reason for such traits up in the air. Scientists who stick to a strictly materialist view of the world look for such things as "non-shared environmental effects," such as prenatal development, as an explanation, but there is no evidence for that; neither does it make much sense.

So, what about those identical twins who grow up differing drastically in things as seemingly fundamental as sexual orientation or proclivity to commit violent crimes? If they have identical genetics and were brought up in virtually identical environments, what would account for these differences?

Most of us are aware of (or may be part of) a family where siblings are radically different from each other—their personalities, interests and value systems sometimes seeming to be completely opposite. It is difficult to chalk this up to either nature or nurture, because both parties couldn't have a more common nature or nurture environment. Having been raised in the same household for their entire lives, and being biologically from the same sets of DNA, what could possibly cause such stark differences?

Mathematically, nature plus nurture doesn't appear to explain why we are the way we are. However, if instead we adopted the well-supported and researched view central to this book, that we are not our bodies, then our consciousness exists independent of our bodies. As such, it is reasonable to expect that that consciousness learns, adapts and evolves across multiple lifetimes, and perhaps even from non-physical experiences. And this would certainly provide an excellent explanation for the anomalies listed above. It would make sense, for example, that IQ, perhaps being related to the function of the brain, be largely influenced by genetics. It would make sense that sexual proclivity may have something to do with your gender in your most recent past life. However, it would not make sense for value systems to be genetic and the influence from family environment would only extend back to childhood; hence, personality traits should show some nurture correlation from "this life," with the majority of the influence being from past lives (and therefore, a mystery to those who don't understand or accept this paradigm).

Therefore, the anomalous nature of significant differences between identical twins raised together can be taken as evidence for a deeper source of influence of traits and values. If it doesn't come from genetics or the environment, it must be elsewhere.

Evidence: The Finely Tuned Universe

Much has been written about the fact that so many constants and laws of nature seem to be perfectly tuned not only for life to have formed, but also for matter to have formed at all.

"The chance that higher life forms might have emerged in this way is comparable to the chance that a tornado sweeping through a junkyard might assemble a Boeing 747 from the materials therein."
- *Fred Hoyle, UK astrophysicist*

Sir Fred Hoyle, distinguished astrophysicist who coined the term "Big Bang," calculated the odds of cellular life evolving in the traditional evolutionary manner to be 1 in $10^{40,000}$.[62]

Consider some of the following instances of perfect tuning in our reality:

- Universal constants cancel out all of the vacuum energy to an amazing accuracy of 1 part in 10^{115}.
- A deviation in the expansion rate of the early universe of 1 part in a billion in either direction would have caused the universe to collapse immediately or fly apart so fast that stars could never have formed.
- A tiny difference in the ratio of the electric field strength to gravitational field strength would have prevented any kinds of molecules to form.
- If the ratio of the masses of protons to electrons were not precisely what it is, chemical reactions could not take place, rendering life impossible.
- The strength of the strong nuclear force could not have differed by more than 2 percent without either preventing hydrogen from being the only element in the universe or from being too rare to allow the formation of stars.
- The number of electrons in the universe must equal the number of protons to an accuracy of 1 part in 1,037 in order for gravity to work.
- A slight difference in the ratio of the number of photons in the universe to the number of baryons would preclude star formation.
- A small deviation in the value of the electromagnetic coupling constant would prevent molecular formation.

Taken together, these facts point to an incredibly perfect set of circumstances for our reality to exist. Creationists would say, "Of course the universe is finely tuned. God created it that way." And they would sleep well at night. Scientists, however, have had to resort to a great deal more thought.

The Anthropic Principle, originally proposed in the 1970s by Brandon Carter, and then significantly developed by John D. Barrow and Frank J. Tipler in the 1980s, goes something like this: "Of course our universe is finely tuned. We wouldn't be here to observe it if it weren't." This idea is a perfect partner to the multiverse theories, for example. Zillions of universes can be formed via your favorite parallel universe theory, all with different physical constants and ratios, leading nearly all of them to a matterless lifeless destiny. But, like that lucky guy who won the billion-dollar lottery by picking the right six numbers, somewhere there is that one in a zillion universe that got all of the physical constants and ratios just right. And where else would we live?

Doesn't Occam's razor apply here? Which theory is more likely? That zillions upon zillions of universes are spawned every second, with every conceivable configuration existing simultaneously, thereby accounting for the anthropic principle? Or, was the universe simply intelligently designed utilizing a technology that should be available to us within this century? Considering that a century is an incredibly small unit of time in the scope of cosmic evolution, it certainly seems plausible—no, almost certain—that this has been done by now. And if so, being a reality generated at a level separate from that which generated it, so must our consciousness exist at another level.

Evidence: Universal Human Spiritual Experience

Earlier in this chapter, I outlined how the concept of reincarnation has an almost universal presence in world cultures across thousands of years. But there are also many other core aspects to Digital Consciousness theory, for which the religions and philosophies from ancient cultures around the globe seem eerily prophetic and consistent. Those aspects include:

1. *All that there is (ATTI, aka God) is the source of everything. We are part of ATTI; hence, we are all interconnected*—The opening three words in the *Sri Guru Granth Sahib (SGGS)*, the holy book of the Sikhs, are *Ek on kar*, which means *"God is one."* This is the very foundation of the Sikh philosophy. Guru Nanak Dev Ji, the founder of the faith, had a mystical experience while he was meditating by a river. People thought that he had drowned because they found his clothes on the riverbank and he was unseen for 3 days. But he returned, having had a revelation that God is one and recognized all of mankind as part of that one. "There is but one all pervading spirit, and it is called the truth, It exists in all creation, and it has no fear, It does not hate, and it is timeless, universal and self-existent! You will come to know it through seeking knowledge and learning." (*SGGS*, p. 1) "But God is not identical with the universe. The latter exists and is contained in Him and not vice versa." Sikh Guru Gobind Singh likens reality to a play that is brought into existence by God. "[In] the figurative abode of God, there are countless regions and universes" (*SGGS*, p. 8). Creation is "His play which He witnesses, and when He rolls up the play, He is His sole Self again" (*SGGS*, p. 8, 292). Note the similarity to the cyclical universe, the bubble universe theory.

2. *Consciousness creates reality*—In Hinduism, *Brahman* is considered to be the ultimate reality of the universe. Its meaning is described in the sacred texts Upanishads as *Sat-cit-ananda*, which translates to being-consciousness-bliss. So, for thousands of years, it has been recognized that consciousness is central to reality; in fact, it is all that there is.

3. *ATTI evolves according to a process of continuous improvement*— *Maya* is temporary in that we reincarnate many times, but we always go back to pure consciousness in between lives. In this lifetime, we are asleep and we don't awaken until we die. The enlightened man is the one who is awakened, which is a cycle of continuous improvement.

4. *The life we live is a simulation of sorts, a virtual reality, which runs within ATTI—Maya* is an ancient Sanskrit word that means "illusion," and is a central theme in both Hinduism and Sikhism. According to the Upanishads, the human experience is interplay between eternal consciousness and Maya or the temporary, illusory world. Indeed, our experience is illusory, a simulation of sorts, and all that we can know that is real, is that experience.

5. *The purpose of life is to evolve our consciousness, and in so doing, evolve ATTI*—In Hindu philosophy, life is viewed as a moral journey, an opportunity to build enough karma to discover the creator. Buddhism teaches that it is a goal of one's life to learn to remove desire and that evolving one's spirit over multiple reincarnations will lead to a return to the creator.

What is important here is that ancient religions developed from very similar learnings and experiences that various mystics, such as Buddha (Buddhism), Mohammed (Islam), Moses (Judaism), Guru Nanak Dev Ji (Sikhism) and Jesus (Christianity) had while in a state of being connected to something divine. Buddha had his while sitting under a Bodhi tree, Mohammed in a cave, Jesus in the desert, Moses on the mountain and Guru Nanak Dev Ji by a river. These mystics were separated by thousands of years in time and thousands of miles, yet still had near-identical insights to the origin and nature of reality, God (ATTI) and the interconnectedness of all things. Those insights all mesh perfectly with Digital Consciousness theory.

Evidence: The Mandela Effect

The Mandela effect has become the de facto name for the idea that something that many people remember from the past is somehow changed or rewritten in our reality. It was named for the former president of South Africa, Nelson Mandela, whom many people recall having died in a South African prison which, history now tells us, is untrue. He died, according to all of the historical artifacts in our reality, of natural causes at the ripe old age of 95. I personally have a vague recollection of hearing some news about his demise in prison, but I can't really place it.

That's the thing about memories; they are completely fallible. According to research, when one remembers something, one is not remembering the original event, but rather the last time that you recalled that particular memory.[63] As such, memories are subject to the "whisper-down-the-lane" syndrome of changing slightly with every recollection. So, my vague Mandela recollection could easily have morphed from a confluence of news reports and "Mandela effect" claims that I have heard over the years.

However, that does not explain why large numbers of people would have the same memory of something entirely fallacious. It happens occasionally and has been happening recently with increasing frequency.

One of the most recent and striking examples of the Mandela effect comes from the 1979 James Bond film *Moonraker*. That movie featured a character named Jaws, a huge henchman with metal teeth played by the late Richard Kiel. In one scene, Jaws' Brazilian cable car crashes and he is helped out of the rubble by Dolly, a bespectacled young blonde woman played by the French actress Blanche Ravalec. There is one of those movie moments that any Bond aficionado will recall, when Jaws first looks at Dolly and grins, baring his mouthful of metal. She looks at him and grins, showing her mouthful of metal braces, and therefore, as the music swells, they fall instantly in love and walk off hand in hand. At least that's the way we all remember it, myself included. The only problem is that if you watch the scene today, Dolly has no braces!

Those 1970s era Bond movies were full of campy moments like this one. It was done to get a chuckle from the audience. In this case, the gag is that despite their drastic differences, having a mouthful of metal in common is enough for them to instantly fall for each other. And so the audience laughs. That was the entire point. But now, the scene simply doesn't even make sense any more.

This is actually a key difference from another well-known Mandela effect regarding the Berenstein Bears (I refuse to spell it any other way), a series of hundreds of children's books popular from the 1960s through the 2000s. Most people recall the name of the series as "Berenstein Bears." But now, all books, and references to them on the Web, are spelled "Berenstain." But in this case, there was no real corroborating evidence that it was ever the way people remember it, other than all of our fallible memories. In contrast, the Dolly, Jaws and braces scenario does have separate corroborating evidence that it was once as we remember it—the very point of the scene itself. In addition, there is a 2014 BBC obituary of Richard Kiel that references the movie, describing Dolly as "a small, pig-tailed blonde with braces."[64] I'm sure the BBC checks their facts fairly carefully and wouldn't typically be subject to mass delusion. In addition, there is at least one image on the Web where Dolly still appears to have braces. So here, it seems, ATTI is really messing with us and didn't even bother to clean up all of the artifacts.

A reality simply cannot change its history and leave its participants recalling something different unless that reality is:

1. Soft, made of something malleable, like bits
2. Consciousness-based

Evidence: Other Anomalies

Hypnosis

(Note: the following paragraph is reproduced from *The Universe—Solved!*)

Author Michael Talbot recounts an event that he witnessed in the 1970s. On that evening, his father had hired a hypnotist to entertain some friends at a social gathering. The hypnotist selected a subject named Tom, whom he had never met before, and put him into a trance, telling him that when he came out of it, his daughter, Laura, would be completely invisible to him. He placed Laura in front of Tom and took him out of his hypnotic state. He stood behind Laura and held a watch against her back so that no one could see the inscription on it. He then asked Tom to read the inscription. Tom leaned forward and squinted as if Laura were not even there, and successfully read the inscription.[65]

We can think about several possible explanations for this event, but all of them take us beyond normal reality. One possibility is that the hypnotist knew the inscription and Tom received it, along with the image of the watch, telepathically during the session. This would require a large assumption on the part of the hypnotist, however, who neither knew Tom nor his capacity for telepathy prior to the session.

If hypnosis can change someone's perception of reality, consciousness would seem to have an active role in the creation of those experiences.

Slime Mold

A slime mold is a bizarre single-celled organism, which has the propensity to aggregate with others of its species to act like a large multicellular organism. Slime molds can be found on your lawn, in your gutters or on decomposing logs, and might reach the size of a square meter or more. In both its unicellular state and in its aggregate slime state, the organism has neither a brain nor a nervous system.

So imagine scientists' surprise to discover that one such representative species, Physarum polycephalum, has shown the ability to learn. Researchers from Toulouse University,

France, placed the mold in a petri dish along with a food source, separated by a distasteful (to the mold) barrier consisting of caffeine or quinine. In the initial run of the experiment, the foul-tasting barrier stopped the mold from getting to its dinner. However, over a few hours, Physarum polycephalum learned to cross over the barrier to get to the food, after which each run of the experiment resulted in faster times and less hesitancy to reach its goal.

This rudimentary learning process requires "a behavioral response to whatever the trigger is, memory of that moment, and future changed behavior based on the memory," which combination would appear to be impossible without a brain or nervous system.[66]

Even more remarkable is Physarum polycephalum's ability to solve complex mazes and emulate Ancient Rome's road-building logic.[67]

As science puzzles over this conundrum, and develops theories based on cellular memory and binary genetic codes, I suggest that the slime mold is simply yet another example of the separation of consciousness and the virtual organism template.

Learning does require a sufficiently complex adaptive system, but that system does not necessarily need to be embodied in a central nervous system of the organism. Quantum mechanics experiments have proven beyond a reasonable doubt that consciousness plays a central role in the creation of reality. This implies that consciousness is not an artifact of the system that it is creating—it is, rather, a separate aspect of reality. Evidence abounds that we live in a consciousness-centric reality and that consciousness is therefore "out there" elsewhere.

Study at Princeton About Clear Days and Group Consciousness Effect

I have been invited to many weddings over the years, perhaps on the order of 30–40. And yet, for all of the weddings that I have attended it has never rained. This is only remarkable because all of these weddings were on the East Coast, where it is common to have summer rain. I've often joked about this anomaly and hinted that casual acquaintances should invite me for good luck.

Of course, there is something silly about this streak, namely that every weekend many people get married over the summer, so it is highly unlikely that the "gods" would ensure good weather for everyone. Has anyone ever done a study to determine if weekend weather is statistically less rainy than during the week?

I have not found such research; however, researchers at Princeton University noticed that Commencement weather was typically always wonderful and so they wondered if a group of people wishing for clear, sunny weather might actually be able to influence the outcome. This is known as the Group Consciousness Effect. There have been many fascinating anecdotal cases of apparent statistically significant effects owing to group consciousness. The Princeton research was one such study.

In comparing average rainfall during 3 key days around Commencement time — P-Rade, Class Day and Commencement — to surrounding days and regions, it appears that there is a statistically significant greater likelihood of good weather during those 3 days, with an error against chance of 2.3 percent. As author Roger D. Nelson, who documented the study for the *Journal of Scientific Exploration*, stated, "A look at actual weather data seems to suggest that precipitation tends to stay away from Princeton for the P-Rade, and Class Day, and Commencement, to a somewhat unlikely degree" and that "if the analysis is correct, the only good candidate to explain the apparent differences, other than chance, would seem to be an influence from an informal but powerful communal wish for dry weather."[68]

Even more significant are the results from the long-running study at Princeton University called the Global Consciousness

Project, which investigates the effects of world consciousness on physical systems. The project consists of an international network of "EGGs," which are computer-based random number generators, running continuously. If the world were truly random, the random numbers would follow a normal distribution or bell curve. And, most of the time, they do. However, at times when the world's attention is synchronized, such as during a catastrophe or international news event, a truly remarkable thing happens. The EGGs display a slight but distinctly non-random pattern, as if some cosmic control center were synchronously influencing the outcome of the generated numbers.

Shared worldwide emotions during events such as Princess Diana's funeral, the Columbine High School massacre, embassy bombings, Y2K and the Winter Olympics seemed to actually impact the mechanized generation of random numbers. In many cases, the effect was noted a short time prior to the actual event itself. Taken as a whole, the data for 209 events "chosen a priori" occurring from 1998 to 2005 departed from randomness by an amount that would be likely to occur by chance 1 in 10,000 times. What could possibly cause this?[69]

As Roger Nelson, director of the Global Consciousness Project, noted:

> We have accumulated a seven-sigma deviation in answer to our basic question: Is there structure in random data during periods of shared attention to global events? The odds against chance are trillions to one, but beyond that, secondary analysis shows further structure. The findings suggest deep unconscious connections among humans that may be the source of correlations we find in otherwise random data.[70]

Evidence to the Contrary?

Just as I presented (and shot holes in) the (very little) evidence that there was for a continuous reality in the last chapter, so shall I present (and shoot holes in) the (very little) evidence that there is for an emergent view of consciousness. Consciousness research in the scientific world consists mostly of observing correlations between activity in the brain and states of consciousness. Neuroscientists attach electrical probes to parts of the brain and, utilizing an oscilloscope, observe electrical patterns in the brain as people have consciousness experience, as the following figure demonstrates:

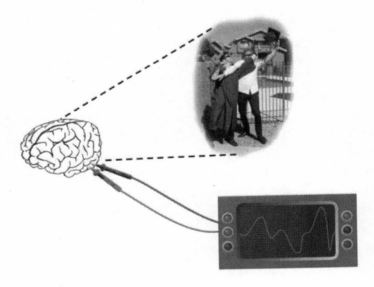

Figure 6.10

They then conclude from this that awareness and conscious experience comes from brain function. In other words, consciousness comes from meat.

And this is all of the evidence that consciousness emerges from the brain. Period. There is no other evidence. So let's see how valid it is as a *hypothesis* for the origin of consciousness.

Imagine if we applied the same logic to a TV set that has been turned on. We apply electrical probes to components of the circuitry of the television and observe the patterns on the oscilloscope.

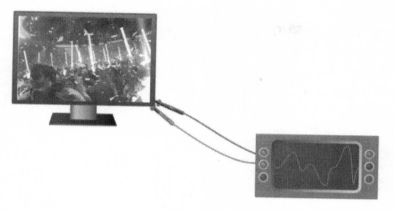

Figure 6.11

Should we conclude that the programming content emerged from the circuitry of the television? Isn't that what dogs and cats think? Dogs and cats have been known to see an image on the TV and walk behind it to see where it is coming from. Aren't we a little more evolved in our knowledge and reasoning than our furry companions? Sure we are, but maybe only because we invented TV and understand how it works. As shown below, the programming content

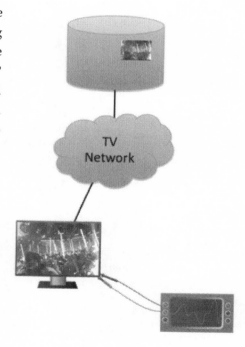

Figure 6.12

exists elsewhere, in the cloud if you will. In the old days, the content was only live in a studio. Later, it was recorded on analog tape. Now, it is digitized and stored on computers.

In a completely analogous manner, there is no reason to assume that our content, our awareness and conscious experience comes from our brains, except that it seems that way, just as the Earth seems flat as you walk around town. But when you travel sufficiently far above the surface of the Earth (a few miles), you can see the curvature of the horizon. Similarly, if you were to travel out of your body, you would realize that your consciousness is elsewhere. In the Digital Consciousness model, the situation looks like the following:

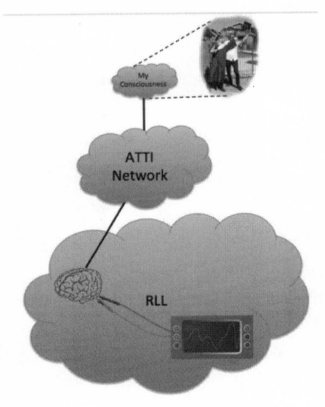

Figure 6.13

The probes that are on your brain are digital templates of probes. Your brain is a digital template of a brain. The signals that are experienced when the probes are attached to the brain are following the laws of the RLL, which is a comprehensive interactive digital simulation of a convincing virtual reality. In this virtual reality, commonly known as *reality*, the appearance of a deterministic objective reality can go arbitrarily deep. Scientists will continue to probe the brain for consciousness, develop hypotheses about how awareness comes about and experimentally discover supporting evidence for it. But, as a result of all of the evidence presented in this chapter, this model of reality doesn't fit. Our consciousness is elsewhere, as shown in the diagram. We interact with other conscious entities through the RLL and have purely subjective experiences.

The Orthodox Church of Logical Objectivism and Willful but Nonsensical Skepticism (CLOWNS)

Merriam-Webster defines religion as "a cause, principle, or system of beliefs held to with ardor and faith." Despite the fact that there is zero empirical evidence for objective materialism (aside from the same evidence that there is for a flat Earth: that it "just seems that way"), and a copious amount of evidence to the contrary, there is a cadre of individuals who believe in objective materialism so fervently that they:

- Disregard all logical arguments against their belief.
- Disparage those who do not believe in their cause with labels like "crackpot" and "pseudoscience."
- Pontificate arrogantly in social media in order to bring more people into their cause.

And, sadly, most of the members of the CLOWNS do all of this without any formal scientific understanding, education or training.

Merriam-Webster also defines "fundamentalism" as "a movement or attitude stressing strict and literal adherence to a set of basic principles."

So, essentially, this group is even more like a high-dogma fundamentalist religion than a moderate open-minded religion. And that's just sad. Because some do have scientific training and should know better. They should know that real science means open-minded skepticism. And that, at any point in time, a real science-minded person should be willing to question centuries of beliefs that no longer fit the prevailing evidence (such as the results of the delayed choice quantum eraser experiments). But, instead, what we have is a group of people who blindly reject anything that is anomalous in that it can't be explained by the dogma of Objective Materialism.

In the 1800s, because their religion couldn't explain rocks falling from the sky, meteoroids must not exist. And still today, because their religion can't explain telepathy, for example, telepathy must not exist. And experiments with statistically significant results that support the existence of telepathy must have been flawed owing to experimenter bias or some other excuse. All of the thousands of such experiments. UFO sightings must all be owing to something mundane, like Venus or swamp gas. Frankly, to people who have a true scientific mentality, blindly assigning Venus or "swamp gas" or "ball lightning" or "mass delusion" to any reported sighting, without a proper investigation, without even speaking to a witness, is laughable and insulting. Really, an object that moves across the sky could be Venus? Really, swamp gas in the desert? Really, bright green ball lightning, followed by the same sized bright blue ball lightning, that all move in a straight line with a constant speed, on a perfectly clear night? Really, 100 people, including lawyers, pilots, engineers and scientists all reporting the same thing is mass delusion? Please.

Yet, for every brave, highly qualified PhD, who goes out on a limb to investigate something that doesn't fit the dogma of the religion of Objective Materialism, you will find, on the Web, both forum posts and blogs full of slanderous accusations. If you don't believe me, pick any name of one of the scientists mentioned (people who have PhDs in a scientific field or MDs in a medical field) in this book, and Google that name along with "pseudoscience" or "crackpot," and you'll see what I mean.

This religion even has secular leaders, akin to the pope, or a bishop, or L. Ron Hubbard—people like Michael Shermer, James Randi and Richard Dawkins are like revered elders of the Church to their followers. But listen to almost any debate between these folks and someone taking the opposing position, and you will often hear name-calling, anger and disrespect on the part of the Church of Logical Objectivism and Willful but Nonsensical Skepticism (CLOWNS), despite a sense of respect, logic and patience from the opposition.

Michael Thalbourne's paper "Science vs. Showmanship: A History of the Randi Hoax" exposes Randi for what he is—a non-scientific showman with a predetermined agenda:

In short, Randi is a showman rather than an unprejudiced critic. Parapsychologists find it difficult to see the merit of working with someone who makes a living out of debunking paranormal claims and who would in fact suffer loss of face and finance if psi came to be more widely accepted as an established scientific fact. (Randi frequently puts the evidential status of ESP on a par with that of unicorns and Santa Claus.) Randi has admitted that he makes a good living out of such debunking through his books and speaking engagements. A second reason is that Randi is violently and implacably hostile towards persons whom he deems to be fraudulent psychics; yet many reasonable persons believe he tends to reach the verdict of "pseudopsychic" much too quickly and on insufficient evidence. Many parapsychologists believe that he is so incapable of suspending his belief that the claimant is not psychic that his attitude will spill over into interaction with the subject and be counterproductive.[71]

To be sure, Randi has no academic credentials, having dropped out of high school at 17, and can't properly argue concepts like statistical significance, the scientific method, forms of logic or the observer effect. He even proudly admits that his Diner's Card states his occupation as "Professional Charlatan." And yet he is the hero to many pseudoscientists and Wikipedia editors.

The pseudoscientists show up frequently on various science forums, such as PhysicsForums.com. Not that these sites aren't chock full of well-educated members of the physics community — they are. But, as there is no test to become a member, nor any educational filter criteria, they are also full of science neophytes, who have locked into the Objective Materialist paradigm, partly owing, perhaps, to poor secondary school education, partly to being raised in a culture of closed-mindedness and partly just wanting to belong to a group. They troll forum threads about alternative ideas, and contribute useless snarky and degrading comments about the subject, without really understanding the topic in the first place.

Scientific experiments supporting the notions of paranormal phenomena are often the subject of attacks, whereby the CLOWNS always claim that such experiments lacked proper "scientific rigor," and then roll out the tried-and-true insults of "crackpottery" and "pseudoscience" on the experimenters. Oddly enough, no one ever bothers to question the quality of experiments that don't challenge the materialistic hypothesis.

Wikipedia has also been completely infected by the CLOWNS. Every article on an alternative thinker has been edited by a member of the Church, usually resulting in sloppy, disjointed articles. Many reputable scientists and researchers have even been denied a Wikipedia page owing to the ignorance of some teenager living in his parents' basement and enamored by the delusions of the Great Randi. Wikipedia contains full-length bios of "Snooki" and "The Situation" from the MTV TV series *Jersey Shore*. But someone who has developed unique and well thought-out solutions to the mysteries of quantum mechanics? Denied. LOL.

I include this admittedly disparaging section on the CLOWNS to expose them for what they are: unscientific, closed-minded frauds whose opinions should simply be ignored. It's my right. It's my book. ☺

Summary

As should be abundantly clear at this point, there is *zero* empirical evidence for materialism or realism but, as this chapter demonstrates, there is a tremendous body of evidence supporting idealism. The evidence comes from an incredibly diverse set of sources; including rigorous scientific experimental results reproduced the world over, evolutionary logic, well-researched NDEs, and corroborated out-of-body and past-life experiences.

Experiences that clearly demonstrate an external seat of consciousness have been common throughout all cultures throughout history. It is only because of a very recent and narrow school of scientific thought (objective materialism) in the Western world that the veracity of such experiences has been suppressed, but even that is beginning to change.

Statistical analysis of the results of psychical research in a wide range of fields (precognition, telepathy, group consciousness) demonstrate to a degree of confidence that is generally considered "fact" in science that consciousness cannot emanate from brain function as an emergent property.

And finally, the raw explanatory power of a consciousness-centric model of reality can no longer be ignored. The next chapters will use this model to explain everything, from every single quantum mechanical anomaly to the nature of the universe, and every single metaphysical anomaly.

Chapter 7

What It Is

"The very study of the external world led to the scientific conclusion that the content of the consciousness is the ultimate universal reality."
- *Eugene P. Wigner*

Here we are again, putting it all together in Chapter 7—perhaps only fitting since numerologists identify the number seven with truth-seeking. Chapter 1 gave a teaser on Digital Consciousness theory. In this chapter, we shall develop those ideas in much greater detail.

There are many models out there of reality, including digital determinism, the holographic projection model, materialism, creationism, the computer simulation model and Digital Consciousness (which is similar to Tom Campbell's Absolute Unbounded Manifold (AUM)). Working backward from the evidence presented in the previous sections, only Digital Consciousness fits. And it is a near-perfect fit and does not need any modification. The value of this and the next chapter, beyond the work that Campbell has already done in his book *My Big TOE*, is to:

1. Further develop the model using easily understood pictorial representations of the key artifacts of reality.
2. Present the solutions to the mysteries of life and the universe that we have been discussing utilizing this model.

The table below presents the logic behind the fit of the evidence to Digital Consciousness theory.

Theory	Supports Digital Evidence	Supports Evidence of Separate Consciousness	Explains Metaphysical Anomalies	In Sync with Mystical Experiences	Explains Quantum Anomalies
Digital Determinism	Yes	No	No	No	No
Holographic Projection	Yes	No	Some	No	No
Materialism	No	No	No	No	No
Creationism	No	No	Some	Some	No
Posthuman Simulation Theory	Yes	Yes	Some	No	No
Digital Consciousness	Yes	Yes	Yes	Yes	Yes

Figure 7.1

Digital determinism, of the type envisioned by Konrad Zuse and Ed Fredkin, like materialism, does not support the concept of a free will. Neither does it provide any explanation for the evidence of a separate consciousness. And it has no mechanism to explain metaphysical or quantum anomalies.

Many scientists and mathematicians have latched on to the idea of the Holographic Principle, whereby our reality may be a giant 3D projection of a two-dimensional surface. This is not to be confused with the general holographic paradigm, which is just an assertion that there are underlying connections between elements of reality, which theoretical physicist David Bohm referred to as the implicate order. That principle is accurate at a higher-level model, where the underlying order is the programmatic construct. But the Holographic Principle that Kip Thorne, Stephen Hawking, and others muse about related to black hole surfaces is a weak idea that makes little sense and doesn't explain much of our real world.

Materialism fails on all counts. It can't provide an explanation or reason for digital and discrete evidence, nor for all of the evidence for the duality of mind and body. Metaphysical anomalies have no place at all in materialism, which is why hard-core materialists are so adamantly opposed to and in denial of any of the evidence for such anomalies. In terms of quantum weirdness (the lack of objective reality and locality), materialists cling to hidden variable theories that have been experimentally discounted to ridiculously low probabilities (1 in 10^{80}).

I included creationism here as a philosophy with limited explanatory power. Although it exceeds materialism on that score, it only does so because it is so easy to say, "God made it so." In reality, it makes no sense in terms of explaining digitalism or the idea of a separate consciousness. It might have some explanatory power when it comes to some metaphysical anomalies such as NDEs, but not for others, such as UFO sightings. And what could creationism possibly have to add to quantum entanglement or the quantum Zeno effect?

The Simulation Theory, à la Nick Bostrom's post-human simulations, does effectively support most of the evidence presented in this book. In fact, although no one has developed the mechanisms particularly well, even some of the metaphysical anomalies can be explained with this model. For example, telepathic experiences could simply be an "out-of-band" data stream between the experiencers playing the sim, not unlike how gamers talk to each other via headsets and a separate data channel while they play the game. However, the Simulation Theory doesn't explain quantum anomalies very well (what would the motivation be for their existences?); nor does it offer much of an explanation for reincarnations, OBEs and "oneness," especially in terms of their meaning.

So, the only theory that explains all categories of evidence for digital reality and a separate consciousness, as well as all quantum and metaphysical anomalies, is Digital Consciousness theory (DCT). But what is it?

Any model of all that there is will suffer from the origin

problem. If you ask the questions why and how enough times, there will always come a point where the answer is "we can't know." It is true for the Big Bang Theory as much as it is true for creationism. But Tom Campbell and Steven Kaufman have independently taken an admirable attempt at working backwards toward an origin theory that is as primal as it can possibly be. And it is elegantly simple, which by itself, as well as the principle of Occam's razor, might earn it points for likelihood. The following summarizes their origin theory:

In The Beginning...

... There was nothing. A great void. And a single simple rule— that things will always evolve toward an improved state. The evidence for this rule of continuous improvement is all around us—evolution in all senses imaginable: spiritual, technological, genetic, social, scientific, etc. And what could be more fundamental as a starting point than a void? So far, so good. Let's represent the void by a simple circle:

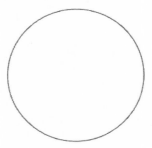

Figure 7.2

At some point, some portion of the void referenced itself against another portion of the void as shown below. Now we no longer have a void, but rather, "something." This is the definition of the most fundamental aspect of space—essentially a binary system consisting of two different things separated in space.

Figure 7.3

If, then, at some point, the portion that was the referencer becomes the referencee and vice versa, we now have the concept of time—the duration between these two events.

Figure 7.4

Our law of continuous improvement will eventually stumble upon the idea that it is more useful to have a periodicity to these events, rather than happening randomly. Now, we have a heartbeat to the system. Further evolution is inevitable, as each of the two sections further subdivides itself. Steven Kaufman developed the concepts of space, energy and matter from fundamental principles in some detail in his book *Unified Reality Theory*. Carrying this forward, we would ultimately have an ATTI that is a digital system, incredibly complex, with a fundamental heartbeat or clock cycle:

Figure 7.5

As mentioned in Chapter 1, ATTI is then a programmatic substrate, upon which we can "program" anything—minds, experiences, thoughts, cars. It has consciousness. It *is* consciousness. Logically, any self-sustaining system that is sufficiently complex would have consciousness. But it doesn't make sense that there would be some measure of complexity, beyond which consciousness exists and below which it doesn't. Rather, it would make sense that the more complex the system is, the more consciousness it has. Working backward, the first subdivision of the void would have been the birth of consciousness; although as Tom Campbell put it, it was incredibly "dim."

What is a System? A Complex Adaptive System?

Merriam-Webster defines a system as "a group of related parts that move or work together." That is a fairly broad definition and can apply to almost anything. System theorists have identified attributes of systems that are specific to certain important categories of systems. For example, a complex adaptive system contains two qualifying adjectives, "complex" and "adaptive." Complex can be taken to mean that the behavior or the system cannot be determined from the analysis of its parts. Adaptive can be taken to mean that the system has feedback loops, is able to modify itself based on some outcomes of the system and, essentially, is able to learn or evolve.

Human beings are certainly complex adaptive systems. As are ecosystems, traffic jams, the stock market, business and the Earth. More importantly, all that there is (ATTI) is a complex adaptive system, containing components like ICs like you and me. Therefore, it evolves and learns just as we do.

Whether or not ATTI evolved exactly in the manner described above, and whether or not this represents just one component in a nested or fractal evolution of reality, doesn't really matter. What matters is that the state that we are in is now a digital system of incomprehensible size and consciousness that continuously evolves toward a more profitable state. In fact, it evolves in at least two ways:

1. As a result of the evolution of the ICs that comprise it.
2. Via higher-level learning mechanisms, such as establishing virtual reality experiences or making adjustments to the overall system based on the results of existing processes.

Another point to observe here is that the system is dispassionate. There is no "agenda" behind its evolution. It merely evolves through trial and error. A change that results in a less profitable or efficient system would be rejected, not unlike frequent saving in a video game, so that in the event that you make a mistake that terminates you, you can go back to the last known stable state. On the other hand, a change that results in a more profitable or efficient system would stick and become the new state of ATTI, ready for the next change. As noted in Chapter 6, some of the most recent experiments in quantum mechanics demonstrate that the system dynamically creates our reality based partly on our conscious needs and partly on pure efficiency. This is exactly what one would expect from a highly advanced complex adaptive system whose purpose is to continually raise the overall level of consciousness.

As an aside, the dispassion of the system allows for the free will of the ICs to generate so-called evil in the world. The notion of evil and, indeed, of good and bad is a human concept, not a fundamental one. That isn't to say that there isn't benevolence in the system. There is, but it is in the form of the consciousness of others, and of their and your intent, more about which later.

"The digital world... consists only of organization—nothing else. Reality is organized bits."
- *Thomas Campbell, physicist*

Boiling the Ocean

As described, ATTI is incredibly evolved and therefore incredibly efficient. Driven by the law of continuous improvement, it has long since learned to make use of any technique that would lend itself to improved efficiency. One such technique is "Divide and Conquer." Imagine that your goal is to become the leader of your country. How might you go about achieving such an objective? Certainly, you wouldn't start knocking on doors in an attempt to convince everyone in your country that you are the best candidate for the job. People generally have ingrained ideas that are difficult to change, so even if you were extremely persuasive, it might take you 4 hours on average to have a 50 percent chance at convincing someone. Working tirelessly 12 hours per day and 7 days per week, it would take about 1,000 years to convince enough people in a country with a population of 100 million, so you had better get started. Clearly a poor plan.

A much better one is to use exponential math to your advantage. Convince two people to go out and each convince another two people, to each go out and convince another two people, and so on. Convincing someone to be your personal evangelist will certainly take longer than getting their vote, but let's say you can do it in a day. With this strategy, you might get your 50 million voters in 52 days. (In reality, of course, political campaigns are organized to take advantage of other effective techniques, like broadcasting, alliances, slander and lying.)

To take another example, let's say that your objective is to raise the temperature of the ocean by 1 degree. You could put a lone 1,000-watt heat lamp on the shore of the ocean. Assuming lossless heat transfer and convection, it would take about 78,000 times the age of the known universe to accomplish the goal. On the other hand, if you put 1,000-watt heat lamps over every square meter of the surface of the ocean, it would only take 200 million years, give or take. However, if you placed a 1,000-watt heat lamp in every cubic meter of the ocean, you could raise the temperature by 1 degree in under an hour.

So, in an analogous manner, ATTI, being efficient, realized that the most effective way for it to raise the level of its own consciousness was to break itself up into small pieces of consciousness and give each piece the motivation to raise its own level of consciousness. Those individual pieces of consciousness are you and me, and the rest of humanity and the kingdom of conscious entities. So now, ATTI looks something like this, where each of the smaller clouds represents the consciousness of you, me, Bill, Hillary and everyone else:

Figure 7.6

But how do you and I raise the level of our own consciousness? By having a learning experience in which we learn how to love, support and provide service to others. You might wonder how we know that love, support and service are the attributes of higher-level consciousness.

Great question. And as with all great questions, the answers are never definitive. But we have some strong clues...

First of all, it is common cultural wisdom across the globe and across millennia that the purpose of life is to learn to love, support and be of service. These beliefs don't come out of thin air. The origins are often owing to mystical experiences, NDEs, OBEs and the learnings that come from those experiences. None of the world's religions or spiritual practices teach that the purpose of life is to grab all that you can before you die. This consistency is significant, because the message of love flies in the face of how we typically live our daily lives: competing for resources, and trying to outsmart and outmuscle our neighboring groups of humans.

Common sense also dictates that the greater consciousness

system evolves through improving these qualities. Recall that ATTI operates under one fundamental rule—that of continuously improving. But what does improving mean? In any system, improvement happens when it becomes more efficient, more productive and better able to achieve its overall objective. Efficiency and productivity often occur when the elements of the system work together collaboratively, with common goals. For example, a company will fail if everyone is working at cross purposes and lacking a common direction. But when everyone is aligned in understanding their purpose and working collaboratively, the company will prosper.

Great things are built through collaboration—societies, scientific advances, improvements to the human condition—but collaboration requires compassion, support and a lack of ego: exactly the qualities that we are talking about when we refer to raising the level of consciousness. So ATTI evolves when its constituents (including us) are aligned to a common purpose, that being supportive, loving and ego free.

But this learning experience couldn't really be done simply by connecting our individuated units of consciousness together and getting them all to have a chat. What would we chat about? How would we help each other? What problems are there to solve? What needs do we each have that others can support? What would drive someone to love someone else?

However, the answers to all of those questions can be found in a reality simulation. By creating an environment whereby we all share common experiences and compete for a finite set of resources, we learn some life lessons that result in the objective—moving the needle slightly in terms of our consciousness quality. This reality simulation is depicted in the diagram as the "Reality Learning Lab" (RLL), as discussed in Chapter 1. Your representation in the RLL is analogous to an avatar or a "template" of a human.

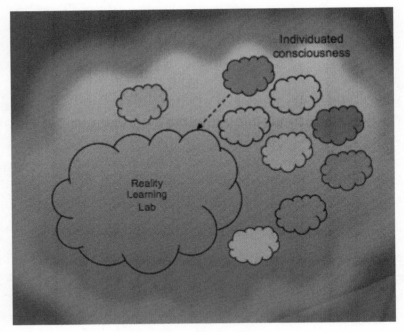

Figure 7.7

In the graphic, the shade of the cloud representing each IC indicates the level of consciousness evolution. You can think of it in terms of "old soul" vs. "young soul." One can easily see that the greater the level of evolution of each IC, the greater the level of evolution of the whole system, ATTI.

If we only had one shot at life, there wouldn't be much opportunity to raise the consciousness level of each individuated segment of the whole. Therefore, it only makes sense that there be some level of continuity of consciousness to build on and that there must be a mechanism for repeated learning through experience in a variety of scenarios. Reincarnation is the logical solution.

Life After Death and the Bigger Picture

As we go through life, and as we go through our "lives," our learnings need to be retained to be useful. So think of them as some sort of database. There are two types of learning databases: one that seems permanent and one that seems temporary. The permanent one is the data storage that carries over from life to life. It is clear that there must be a mechanism for storing such data, because of the evidence for the duality of mind and body presented in Chapter 6. However, does everything need to be remembered? No—in fact, it would be detrimental to the overall goal of continuous improvement if it were. If we recalled everything from a previous life, we would realize that life is temporary, and we might not treat its situations with the gravitas necessary to make the proper decisions and obtain the desired learning. If you were playing a video game, knowing that you can always start over, it would be easy to be reckless with your decisions—just for fun or to see what happened. The consequences of restarting might be insignificant to the fun you are having, wreaking havoc in your environment.

However, if you were playing a fully immersive and realistic video game, and had a mechanism for temporarily suspending your memories, you wouldn't realize it was a game that you were playing, so you would treat every situation and every decision with the seriousness that is needed to obtain the learning. So it is with reincarnation. We retain the values that we have enhanced life to life, but forget the details of the particular life that we had. Partially because it doesn't matter any more and partially because we need to maintain the illusion of the physical reality in order to learn. And yet, those details are never fully erased, as evidenced by the information that people sometimes pull out of a past-life regression or a deep meditative state.

So here's how it all works. As shown in Figure 7.8, as we go through life, we are connected to the RLL and collect our learnings in two places: one that I refer to as sticky soul data and the other being the store of data that won't matter in the next life:

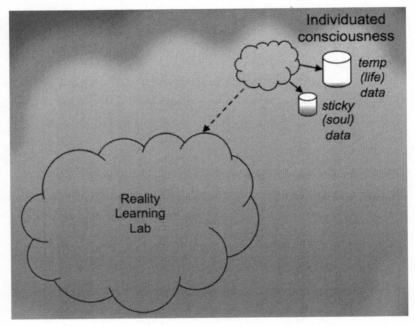

Figure 7.8

When we die, we drop the connection to the RLL and exist in a pure state of consciousness. This is why people who have detached from their waking reality during an NDE experience something otherworldly, where "time disappears," because time and space are artifacts of the RLL only.

The temporary life data is reset. This isn't to say that it disappears entirely. As mentioned above, ATTI appears to keep a record of everything and, in certain situations, some of it is accessible. Such experiences are accessing this record stored somewhere in ATTI. However, in general, the connection to the life data, no longer being needed, disappears when you die, or rather, after you have had your life review.

Then, according to the collective experiences of those who are lucky enough to be able to travel outside of the RLL during their lives (see OBEs), you hang around in a very different environment between lives. Campbell calls this region a non-physical matter reality (NPMR), because it is not an *apparent* physical reality but is indeed a reality inasmuch as it is fully experienced by your consciousness. Here is where you interact with other entities: people who have passed, people who like to help others design their next life, other people like you. With the help of a spirit guide or two, you work out the plan for your next life: what you intend to learn and the basic circumstances of your life. You never lose track of your sticky soul data, as it represents all of the evolution of your consciousness since its inception.

When you are ready, you will begin your next life by simply reconnecting to the RLL, where your data stream interacts with an avatar that is a baby or a fetus belonging to the family of your choosing. You get a blank slate of temporary life data to start filling up with memories, but you retain your soul data. This may explain why identical twins can have such drastically different personalities and values, or why some children have a natural affinity for music, water, cats, sports or whatever.

Note that the clouds and little database icons are just models. The actual underlying mechanisms are unknown and perhaps even unknowable. But the model is very effective in explaining what appears to happen as you live, die and reincarnate.

"Life isn't as serious as the mind makes it out to be."
- *Eckhart Tolle*

What is Real? What is Virtual?

It is fairly common to feel that our reality is the most fundamental level of existence. It feels physical, hard, and with a high degree of consensus of experience between us and our peer entities (other humans). But, as demonstrated in Chapter 6, this is all an illusion. Our waking reality is not that hard and is actually virtual. The degree of consensus is high, but not 100 percent. The illusion looks like the levels shown in Figure 7.9. Our outermost level of reality is our waking world — an apparently "physical" reality. I have arbitrarily assigned this the number 0, so that it makes it easy to refer to. The next level within a particular level is identified with a number that is one less. The level numbers, of course, are meaningless and are simply a way of relating one level to another.

When we dream, we feel like the dream is a virtual experience that takes place within our waking reality, but at a level deeper, as in "I fell deeply into a dream state." It is the same with strapping on a VR headset and experiencing a virtual reality in a computer simulation. These "virtual" experiences can be thought of as one level removed, or one level within, the waking or apparent physical level, shown here as -1. And if we have a dream within a dream, or experience a computer simulation within a dream, that conscious experience is one more level removed (-2), "nested" within the reality layer around it.

If you wake from the dream within the dream, you think "Ah, OK, it was just a dream," but that, too, is a dream and when you wake from *it*, you think "Aha, I had a dream within a dream." Each level of awakening effectively pops you out of another level toward waking reality. It is actually completely arbitrary whether or not I put the most fundamental level at the center of the model or at the outermost ring, or level. It is just a metaphor. If you think in terms of "deeper" meaning a more fundamental or "closer to physical" level of reality, you can reverse the order of the elements; it doesn't matter.

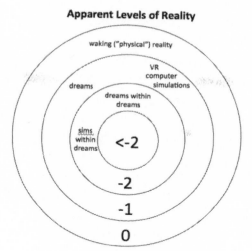

Figure 7.9

However, we now have all of the accumulated evidence previously presented, and the picture of our world is somewhat different. In the Digital Consciousness model, the layers are probably more like what is shown below in Figure 7.10:

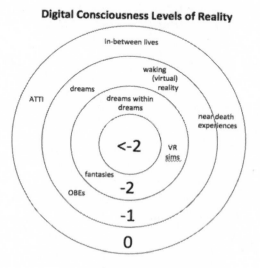

Figure 7.10

We can assume that ATTI is the fundamental construct of reality (0); although, as with the use of the word "universe," we really have no way of knowing for certain at this point. Still, it can be a good working assumption. Within ATTI is our waking reality (-1) which, again, is actually virtual per the connection to the RLL. Interestingly, dreams may not really be a nested experience within the waking reality, but rather a sideways one. As you fall asleep, does it sometimes feel like you are slowly moving into another reality space versus developing a fantasy within your waking reality? OBEs would also fall into this level, as the environment tends to be the RLL or dreamlike.

But a more fundamental, or "closer to the true reality," level can be found in the afterlife or, more accurately, described as "in-between lives." Based on reports from those who can recall this state, it is timeless, unifying and consciousness-centric, like ATTI, and so I include it within level 0. NDEs put us in touch with this domain; although they generally include the experience of facing the "point of no return," which may be thought of as the boundary between level -1 and level 0.

Daydreaming fantasies and VR computer simulations seem to be logically nested within our waking reality at Level -2.

Lions and Tigers and Bears

Why do we have elephants, tigers, dogs, cats, gnats and mosquitoes? I always find it amusing when an article goes viral about the consciousness of animals. *Scientists declare that dolphins have consciousness* shouts a headline.[1] Really? How about *Mystics declare that scientists have consciousness*?

What are the hallmarks of consciousness? It depends on whom you ask. A 2012 paper by A. B. Butler, published in the medical journal *Advances in Experimental Medicine and Biology* states that the characteristics of consciousness are high cognitive abilities, verbal abilities, episodic memories and a sense of self.[2]

Really? I find this definition to be very naïve and human-centric. Because an animal doesn't verbalize, it isn't conscious? Were we not conscious before we developed languages? High cognition? I know quite a few politicians and sports announcers who would fail that one. Honestly, we have no more business putting animals into "conscious" and "non-conscious" categories than we have classifying humans into "worthy of voting" and "not worthy of voting" categories.

For a different take, ask someone more enlightened. Dr. Ervin Laszlo is a Hungarian philosopher of science who has written many papers and books on science, consciousness and Eastern philosophy. He identifies consciousness with characteristics such as the recognition that one is part of nature, behaving in a cooperative and non-competitive manner, and appreciating one's connectedness with everything else.[3] By that reasoning, ants and bees might have more consciousness than humans, which also doesn't feel right.

Simply put, consciousness is awareness and self-reflection. Anyone who spends time around animals will recognize those characteristics easily. Does that mean that there is an IC in ATTI associated with each elephant, tiger, dog, cat, gnat and mosquito in the RLL?

Why not? Is there a fundamental reason to assume that just because something is small relative to us, that it isn't conscious? Maybe the same reason that we used to think the universe revolved around the Earth...

On the contrary, it strikes me that there are some very good reasons to assume that all life forms are conscious (in general; see below for caveats):

1. There is no evidence to the contrary.
2. As time progresses, even scientists acknowledge that one species after another is conscious. The trend appears to be that they will continue to accept consciousness in animal

species, one by one, even as the species under study has smaller and smaller brains, and therefore, ostensibly, less and less consciousness.

3. A rich variety of conscious species provides a rich learning experience in our lives.

I submit that there is not a lower limit to consciousness; that it is not an emergent property that suddenly appears upon some level of mental complexity exceeding a tipping point. Therefore, all living species are conscious to some level. It is just that some are "more conscious" than others. If so, there may be another, more esoteric, explanation for the variety of species. In ATTI's effort to self-organize to be as efficient as possible, it may have discovered that it is most efficient to have a variety of "consciousness" sizes packed together in the whole. Just as a pile of boulders leaves room between them for rocks, and a pile of boulders and rocks has gaps for pebbles, and a pile of boulders, rocks and pebbles still has gaps for sand, so it may be with consciousness, as the figure below demonstrates:

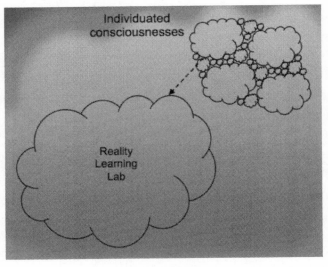

Figure 7.11

I Am a Rock

"Life is a whim of several billion cells to be you for a while."
- *Groucho Marx*

So a human has consciousness. And a dolphin. And dogs and cats. And gnats. What about plants? Plant communication researcher Cleve Backster famously (or infamously, depending on your level of skepticism) performed experiments on plants that seemed to indicate consciousness. Using galvanometers (the same device used to measure skin resistance in lie detectors, a technology in which he was an expert), he measured the "emotional" response of plants to various stimuli, including the intent to harm the plant, which resulted in a strong reaction from the plant. Further experiments demonstrated that the plant was able to "recognize" an individual who had previously destroyed another plant.[4] The results of these experiments received a lot of popular press and attention, and of course, as a result, a strong backlash from the scientific and skeptic communities. Fair enough, as Backster's tests were not rigorous, didn't use control groups and weren't particularly repeatable. However, over the subsequent decades, a variety of experiments demonstrated plant awareness and communication.

In 2008, for example, scientists from the National Center for Atmospheric Research discovered that walnut trees produce an aspirin-like compound in response to various forms of stress. This seems to indicate some level of self-awareness. Then, in 2012, scientists from the University of Exeter, UK, showed how plants can warn neighboring plants of imminent danger.[5] And the Royal Horticultural Society Centre did a recent month-long study on the effects of talking to plants. Using a recording of 10 different people (5 male, 5 female) reciting scientific or literary works, it was found that plants subjected to female voices grew statistically significantly taller than either the control group or the group subjected to men's voices. In fact, it was the great-great-granddaughter of Charles Darwin whose voicing of a passage from *On the Origin of Species* yielded the best effect. I wonder what Chuck would have to say?[6]

A current and popular argument against the idea of plant consciousness relates to the fact that plants do not have a nervous system. However, as discussed in Chapter 6, there is plenty of evidence that at least some aspects of consciousness reside outside the brain anyway, which tends to defeat the argument against the likelihood of plant consciousness.

Again, in the Digital Consciousness model, it is not the complexity of the nervous system that makes a conscious entity. Rather, it is the *decision* of an IC within ATTI to make use of a vehicle, such as the template of a particular living entity, as an experiential substrate. Through that template the IC can experience waking reality and evolve via that learning. In order to actually have the experience and to interact with the environment in a meaningful way, the template must be sufficiently complex to support such interaction. So the important question is actually different: Does a plant have sufficient complexity to support interaction, the collection of data from its environment and the generation of a useful response to the conscious intent of the IC?

This is a far less restrictive set of criteria. Note that it is not necessary that the plant have the structure to be able to "think." That can be left to the consciousness. In fact, recent research supports the idea that plants have complex sensory and communication mechanisms, memory and a non-neuronal yet sufficiently developed cognitive structure.[7] So, yes, a plant is mostly likely conscious. Or slightly more accurately, and as with animals, a plant is a vehicle that supports consciousness.

What about inanimate objects? Do rocks have consciousness? On the face of it, we might be inclined to say, "Of course not; rocks aren't alive!" But notice that that is a different question. We all kind of know what we mean by "alive." A dog is, a rock isn't, a tree is, a cloud isn't. "Alive" implies some sort of autonomy and ability to respond to stimuli. Reproducing a slightly modified section from *The Universe—Solved!*, below, demonstrates how difficult it can be to define life:

The Definition of Life

Think you know what life is? Guess again. Conventional biologists typically define life as anything that generally exhibits all of the following properties:

1. Growth
2. Metabolism
3. Motion
4. Reproduction
5. Homeostasis (or responding to stimuli)

However, it turns out to be surprisingly difficult to consistently define life according to these or any other rules. For example, we would all agree that a mule or a drone bee is alive; although, being sterile, they violate property 4 above. Viruses have all of the above properties and are sometimes considered to be living organisms, but not by certain biologists (aka "cell bigots"), who believe that only cellular organisms can be considered alive. Fire and crystals have all five life-defining properties, but most of us would not consider them to be alive. What about the Earth? It certainly has the properties listed above, with the exception of reproduction. If one can consider humans nothing more than a large colony of living cells, then isn't the Earth just a larger such colony? What about an autonomous robot? Or the upcoming marriage of artificial intelligence and nanotech?

But what if a consciousness was able to occupy a cloud and experience, or feel, what it is like to be a cloud; to move, float, spread out and dissipate? Given that these are experiences that we can imagine having, then it seems possible that consciousness could "occupy" a cloud.

In the 1800s, physicist James Clerk Maxwell found it necessary to add a new component, the electromagnetic force, to the list of building blocks of physics. Similarly, cognitive scientist and philosopher David Chalmers recognizes that we need to add consciousness to today's list of fundamental building blocks of reality. He further speculates that consciousness, being fundamental in this way, therefore "exists in all things—from human beings down to photons."[8]

A related concept, panpsychism, is the idea that everything contains some amount of consciousness, including the smallest component of matter. Panpsychism actually has a rich history of development and a respectable legacy of adherents, such as Plato, Baruch Spinoza and Gottfried Leibniz. It fell out of favor in recent centuries because of the (largely Western) scientific focus on materialism, but open-minded philosophers and scientists recognize that with what Chalmers coined "the hard problem of consciousness" (why we have phenomenal experiences at all), panpsychism at least offers a framework for a working explanation. Christof Koch, chief scientific officer at the Allen Institute for Brain Science in Seattle, and member of Scientific American Mind's board of advisers, notes, "Consciousness comes with organized chunks of matter. It is immanent in the organization of the system."[9]

But why would a "live" brain have consciousness and a "dead" one not, if they each have the same level of complexity? To answer that, psychiatrist and neuroscientist Giulio Tononi of the University of Wisconsin–Madison developed a framework that he calls Integrated Information Theory (IIT). This theory suggests that consciousness is more than complexity and organization; it also requires highly integrated states of being. Because a sequence of memories or experiences are correlated, or integrated, they have meaning relative to each other and this defines consciousness.[10] Clearly, a dead brain cannot correlate experiences and, hence, not support consciousness.

This model is interesting and certainly more insightful than simply saying that our consciousness emerges from the complexity of our brains, but it is still derived from that assumption which, as discussed at the end of Chapter 6, is an assumption with no real evidence. On the contrary, the hard problem of consciousness simply disappears once it is recognized that consciousness is separate from the brain.

A separate but IC in ATTI connects to a chosen template of a living entity in the RLL, which has sufficient complexity and capacity for integrated experiences. And that template may be one of what we would normally consider inanimate, such as a cloud. The consciousness experience that we might have with such a vehicle may be very dim indeed, but it is still possible. So, yes, a rock may have a very limited level of consciousness.

None of these ideas is particularly foreign to those who have been immersed in Eastern cultures. There is a Sufi saying, for example, that, "God sleeps in the rock, dreams in the plant, stirs in the animal, and awakens in man." The Chinese Buddhist Encyclopedia states that, "in Tibetan Buddhism and Japanese Buddhism, all beings (including plant life and even inanimate objects or entities considered 'spiritual' or 'metaphysical' by conventional Western thought) are or may be considered sentient beings."[11] According to Indologist Wendy Doniger O'Flaherty, the Indian concept of Maya is that the universe is not an illusion in the sense of being unreal, but rather, "it is not what it seems to be, that it is something constantly being made."[12] This is extremely interesting and coincidental in light of our evidence that consciousness creates reality.

Why do I Need this Gray Matter?

As shown in Chapter 6, the gray matter is part of the grand illusion of objective reality. It is just a component of the digital template of our bodies. So why do we need it? If the conscious processing is being done in our IC, and not in the brain, what is the purpose of the brain? Let me explain by analogy...

Remember that our world is virtual. It may be impossible to ever know when we have identified a real physical underlying construct. Be that as it may, our consciousness is in a more fundamental layer of reality than is our waking (apparent) reality, as described in Chapter 6. It is similar to when you imagine a fantasy in your head—that is being done at the next higher level of virtuality.

So by analogy, we might consider playing a virtual reality game. In this game we have an avatar and we need to interact with other players. Imagine that a key aspect of the game is the ability to throw a spear at a monster or to shoot an enemy. In our physical reality, we would need an arm and a hand to be able to carry out that activity. But in the game, it is technically not required. Our avatar could be arm-less and when we have the need to throw something, we simply press a key sequence on the keyboard. A spear magically appears and gets hurled in the direction of the monster. Just as we don't need a brain to be aware in our waking reality (because our consciousness is separate from RLL), we don't need an arm to project a spear toward an enemy in the VR game.

On the other hand, having the arm on the avatar adds a great deal to the experience. For one thing, it adds complexity and meaning to the game. Pressing a key sequence doesn't have a lot of variability and it certainly doesn't provide the player with much control. The ability to hit the target could be very precise, such as in the case where you simply point at the target and hit the key sequence. This is boring, requires little skill and ultimately provides no opportunity to develop a skill. However, the precision of your attack could be dependent on a random number generator, which adds complexity and variability to the game, but still doesn't provide any opportunity to improve. Alternatively, the precision of the attack could depend on some other nuance of the game, like secondary key sequences, or timing of key sequences, which, although providing the opportunity to develop a skill, have nothing to do with a consistent approach to throwing something. So, it is much better for your avatar to have an arm. In addition, this simply models the reality that you know and people are comfortable with things that are familiar.

So it is with our brains. In our virtual world, the digital template that is our brain is incapable of doing anything in the "simulation" that it isn't designed to do. The digital simulation that is the RLL must follow the rules of RLL physics much the way a "physics engine" provides the rules of RLL physics for a computer game. And these rules extend to brain function. Imagine if, in the 21st century, we had no scientific explanation for how we process sensory input or make mental decisions, because there was no brain in our bodies. Would that be a "reality" that we could believe in? So, in our level of reality that we call waking reality, we need a brain.

Skynet Lives

"The Skynet Funding Bill is passed. The system goes online August 4th, 1997. Human decisions are removed from strategic defense. Skynet begins to learn at a geometric rate. It becomes self-aware at 2:14 a.m. Eastern time, August 29th. In a panic, they try to pull the plug."
- *The Terminator, Terminator 2: Judgment Day*

Could our consciousness occupy a computer system? The idea of a self-aware artificial intelligence (AI) has been a popular subject of many dystopian Hollywood film franchises (*The Matrix, Terminator, Transformers*). But we need to make a distinction between an AI that develops an emergent consciousness and an AI that is fueled by an IC from ATTI, because it is the specter of the former that scares the bejesus out of everyone. So, let's examine that...

I argued in *The Universe—Solved!* that awareness and consciousness are not properties that should emerge from the complexity of a system. A simple thought experiment involving a callosotomy (where the signaling highway between the two hemispheres of the brain is cut) and a couple of transplants seems to provide logical grounds for doubting the idea of emergence.

Imagine a human subject, Nick, and two lesser primates, Magilla and Kong. We remove Nick's brain and attach it to Magilla's body. Nick should retain his memories and consciousness, but feel really different, since his sensory input is completely new. We would have to conclude that he maintained a continuous, albeit altered, stream of identity. However, given that callosotomy experiments have confirmed that each half of a brain can carry on supporting the conscious existence of its host, we could theoretically put half of Nick's brain into Magilla and the other half in Kong. Where is Nick's identity now? Which body does the old Nick feel that he is in?

If we took the emergent consciousness (the mind is only in the brain) point of view, we would have to say that his consciousness is in both primates. That must be very confusing, receiving two separate sets of sensory stimuli and two distinct developing sets of new memories. Given that the state of the two primates is fairly consistent with the state of two similar natural primates, namely that they each have a brain of 50 percent neural complexity, why should there be a single conscious identity occupying both bodies in the case of Kong and Magilla, but two distinct identities in the natural case? The answer is simple, invoking Occam's razor. Nick's soul simply chose which primate to move into, along with his brain. Alternately, his soul could have said, "This is ridiculous. I'm returning to the spirit domain. Let some other souls fight over those abominations."

A writer on consciousness and Buddhism, Sean Robsville, makes an excellent argument against the idea of emergent consciousness in his "Transcultural Buddhism" blog.[13] He notes that the definition of emergence from the *Dictionary of Philosophy of Mind* is "Emergence—Properties of a complex physical system are emergent just in case they are neither (i) properties had by any parts of the system taken in isolation nor (ii) resultant of a mere summation of properties of parts of the system." Upon deep inspection, things that are considered to be emergent, such as the moving and oscillating automatons in Conway's Game of Life, are not emergent in an absolute sense, but only appear to be emergent. Such properties are "algorithmically compressible, with no remainder, back into the rules that generated them. There is no mysterious addition of procedural complexity." Therefore, "if the phenomenon hasn't emerged from the object, then the only other place from which it could have emerged is the mind of the observer. We are therefore left with the conclusion that emergence is a psychological, not a physical phenomenon."[14] David Chalmers comes to the same conclusion in his "Thoughts on Emergence," when he observes:

> The notion of reduction is intimately tied to the ease of understanding one level in terms of another. Emergent properties are usually properties that are more easily understood in their own right than in terms of properties at a lower level. This suggests an important observation: Emergence is a psychological property. It is not a metaphysical absolute. Properties are classed as "emergent" based at least in part on (1) the interestingness to a given observer of the high-level property at hand; and (2) the difficulty of an observer's deducing the high-level property from low-level properties.[15]

So, if consciousness is a psychologically emergent property, as it appears, then the mind, having that emergent perception, must be elsewhere. And since the mind can't be elsewhere in an AI, it must therefore follow that an AI cannot develop consciousness and awareness emergently.

But what about the second option, which I mentioned in the beginning of this section: an AI that is fueled by an IC from ATTI? If the human body (whether it be embryo, fetus or baby) is a digital template in the RLL, that is occupied by choice of the IC elsewhere in ATTI, why couldn't that IC occupy a digital template that is a silicon-based computer system in RLL?

First, we must note that the digital human template is digital all the way down, much deeper than the quarks, other hadrons or leptons that we have modeled. It is therefore exceedingly complex, from an information capacity standpoint. But our consciousness doesn't care about the digital template of particles. It only requires sufficient complexity of neurons and other body constituents to receive information about the environment, interact with it and perhaps cache some memories.

Applying that same logic to a computer, it would have to have sufficient interconnection complexity to support learning (like a neural net), as well as other transducer components, to receive information about the environment, interact with it and perhaps cache some memories. Once an AI gets to that point, then, it could theoretically host a consciousness, but only if the IC finds value in occupying such a template. Given that the purpose of a consciousness is to evolve, we might ask the question: Would occupying a digital silicon-based vehicle give the conscious entity any kind of useful experience from which it could learn?

I would be inclined to say that we couldn't rule out such a possibility. Hence, a sufficiently complex AI could host consciousness. That said, it is quite a bit different than the *Terminator* (Skynet) scenario, where a learning AI runs amok and takes a preemptive strike at humanity. Why would the same type of IC that occupies our carbon-based bodies in order to improve its quality occupy a silicon-based AI and run amok? As Spock from *Star Trek* would say, it is not logical. We needn't fear Skynet.

That isn't to say that we needn't fear a learning machine that doesn't have the proper safeguards programmed into it to avoid causing a catastrophe. In 2015, a robot at a Volkswagen plant in Germany killed an autoworker. But, its behavior was entirely under the control of the program that humans put into it and it followed that program faithfully. It was unfortunate that the programmers did not foresee every single possible situation or environment and program safely around them. Then again, such a task is impossible. There are schedule and budgetary constraints that make it impossible to create a commercially viable product without sacrificing some "long tail" (rarely encountered) reliability, quality and safety considerations.

Imagine a software system that is controlling a device that has access to the triggering mechanism on a nuclear missile. Sounds like a bad idea to me. Software bugs are inevitable in any complex software system and worst-case scenarios will have a finite probability. But, no matter what, the root cause of any catastrophe resulting from the decisions or actions of a computational system will be a human error. That is because such systems are inherently deterministic.

Neural nets are software structures that consist of logical networks of computational components, which include feedback paths and weighted parameters. The parameters can be tuned based on the outputs of the network, which makes them adaptable and therefore able to "learn" in a rudimentary sense. The interesting thing about neural nets is that, if designed properly, they can learn and modify their own parameters and structure to solve a broad class of problems, without requiring a predetermined brute-force algorithm to do so. As such, they tend to model the neural structures of the brain.

Applications of neural nets include speech recognition, computer vision, financial trading systems, game learning and medical diagnostic systems. Because they are complex and adaptive, it is not obvious from an analytical standpoint how they arrive at their solution. You kind of wind them up and let them go by setting initial parameters, and letting them modify those parameters as they learn about the problem at hand. Yet, despite that seemingly non-deterministic behavior, neural nets are still 100 percent deterministic. Given enough analysis, one can derive the ultimate output from the exact initial conditions, plus the state of all environmental inputs, over time. Like all other AIs, they have no free will, no awareness and no consciousness.

But the fact that we can't predict how a neural net or other complex adaptive system will evolve is definitely disconcerting. When such technology is applied to systems that can have control over weapons, or our financial or social infrastructure (for example, energy grid), extreme precautions need to be taken. That said, and for the reasons outlined above, I don't feel that we need to fear the specter of an AI deciding on its own to wipe out humanity.

Are There Zombies?

By the logic in the previous section, there could be apparently autonomous living entities, which are avatars in the RLL, but which have not been connected to by an IC in ATTI. Such an entity would certainly have a place in the digital consciousness model, and I am sure that most of us have wondered if this might not explain entities who appear to have zero compassion and empathy, like giant Asian hornets, mosquitos and Dick Cheney.

But is there evidence for zombies?

Drs. John L. Merritt and J. Lawrence Merritt II argue in their 2010 book *When Does Human Life Begin* that the scientific, scriptural and historical evidence supports the idea that human life occurs at implantation (about 8 days after conception). However, as should be clear by now, this is not the same as ensoulment, since the definition of life is completely different than the definition of ensoulment.[16] And, unfortunately, science can tell us nothing regarding evidence for ensoulment, because the whole idea of a separate soul from the body is anathema to science.

So we have to dig deeper.

Paramahansa Yogananda, a well respected Indian yogi, and author of *Autobiography of a Yogi*, says that the soul enters the body at the moment of conception. "When the sperm and ovum unite, there is a flash of light in the astral world. Souls there that are ready to be reborn, if their vibration matches that of the flash of light, rush to get in."[17]

In Ancient Greece, Aristotle taught that the human fetus originally has a vegetable soul and that ensoulment occurs at 40 days for males and 90 days for females. This belief was prevalent throughout the ancient Western world from early Christian teachings, including those of Saint Thomas Aquinas, and through the 19th century.[18] Officially however, the Catholic Church has never stated a position on the period of ensoulment, despite condemning the termination of a pregnancy at any point in time.

Some interpretations of the Jewish Talmud hold that ensoulment occurs at 40 days, while others place the event at birth or even when the child first says, "Amen."

According to Badawy A. B. Khitamy of the National Committee for Bioethics, Oman, the Qur'an teaches that the ensoulment period is about 120 days after conception.[19] Other Islamic interpretations differ. For example, the Qur'an also states that, "We made out of the 'embryo' bones, and clothed the bones in 'muscles'." (23:14). Therefore, "the beginning of a human person as an individual living organism is when the embryo develops into fetus at around the 9th week of development (after the 57th day) after the bones and muscles form, but before the development of hearing and sight. Only at this point do we have a multicellular organism and not merely a mass of living cells stuck together. The soul requires that there is an individuated matter present and prior to this period, that there did not previously exist an individual human organism."[20]

These are all ancient teachings, however. And while there may be a valid source of truth in them perhaps owing to a bona fide mystical experience by a prophet who tapped into true ATTI wisdom back in the day, because the traditions are so ancient it is extremely difficult to establish such a definitive connection. So, perhaps the best evidence for ensoulment would come from the correlations of the knowledge attained through modern-day mystical experiences of realms outside of our RLL. In his book *Far Journeys*, Robert Monroe tells of his encounter in the non-physical realm with a pair of entities he called "AA" and "BB," neither of whom had had any experience in our "physical reality." AA, however, became fascinated with the idea of earth as a "school for compressed learning" and went through the protocol to become incarnated as a boy in a New York tenement. His ensoulment occurred within the womb.[21]

So, taken at face value, the collective anecdotal evidence supports the idea that souls, or ICs, choose to occupy digital human templates, or bodies, at some point very early in the life of the human. That means that there is a point in time of human development before which the body has no soul. It seems possible, therefore, that a given human body could be an autonomous robot, following the finite state machine of its digital template in the RLL. Also known as a zombie. But still, there is no real evidence that they exist. Except during election year.

Summary of Digital Consciousness Theory (DCT)

An analysis of dozens of different categories of evidence leads to the inescapable conclusion that our reality is fundamentally different than it seems. This shouldn't be particularly surprising. There are many experiences that we have that belie a deeper reality that seems counterintuitive until we are able to inspect our world from a different scale, often through the use of advanced technology. The sun appears to rotate around the Earth, but it wasn't until advances in observational science that that apparent reality was shown to be false. Walking around town, the Earth seems flat. But take a modern aircraft high enough into the sky, or captain a vessel around the world, and we realize its true shape.

When we look at solid object, we are unaware of the deep and complex molecular structure, unless we have a sophisticated enough microscope with which to inspect it. From the perspective of a stable point in space, the velocity of one moving object plus the velocity of another object moving relative to the first would logically be the sum of the two velocities. But Einstein showed that this is not true and that it breaks down dramatically as the velocities approach the speed of light. And quantum physics has shown us that our apparently objective reality is not objective at all and, rather, depends on the observational state of nearby conscious entities. So is it that surprising to learn that our reality is digital and not continuous? And that consciousness is separate from the brain and is at least somewhat responsible for creating our reality?

This is what the evidence tells us. The implications are significant and address some of the most timeless and fundamental questions in history...

- There is a greater all-encompassing reality, which most closely resembles concepts like "all that there is" (ATTI), God or "universal consciousness."

- We are living in an illusory reality, fundamentally no different than a dream or a fantasy, except for the forced consensus between other conscious entities having the same experience.
- Our consciousness is immortal, as are those of our fellow humans and other living species.
- We have free will.
- We are all connected.
- Our lives have purpose. We choose to have certain learnings in our lives, so as to continually purify our soul. ATTI and the other consciousnesses work together to provide those opportunities.
- We can shape our reality.
- Material things, for which we compete, actually have no meaning or lasting value.
- The ancient mystics were right. Zen, Buddhism, the core spiritual teachings of Sikhism, Sufism, Hinduism, Christianity, Native American and shamanic beliefs were on the mark—we are part of a greater whole, we are all interconnected, we control our destiny, the world is illusory, love is the way and materialism sucks.

Chapter 8

Everything Explained

What if I told you that you could have the answers to all of the philosophical mysteries of the world—what life is all about, what happens when you die and how it all works—would that be worth $19.95?

Don't answer.

Because there's even *more*!

In this section, I will use the Digital Consciousness model to explain all of those pesky scientific and metaphysical anomalies—nature vs. nurture, the nature of matter, parallel universes, dark matter, the observer effect, the placebo effect, quantum entanglement, the quantum Zeno effect, why the universe appears finely tuned, quantum retrocausality, classical retrocausality, telepathy, precognition and UFO sightings. The only thing this model can't explain is the unexplainable. Like Donald Trump's hair.

David Bohm and other holographic theorists use the concept of the "implicate order" to refer to as the basis from which reality emerges. It looks something like Figure 8.1 following:

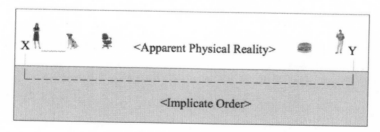

Figure 8.1

The figure above shows two levels of reality. The white (top) level is what we are used to—people, dogs, furniture, hamburgers, stars, bad pop music, crappy politics. Bohm and others postulated a deeper unseen structure of reality, shown here as the implicate order, shaded in gray, which explains certain anomalies and provides the foundation for esoteric experiences.

Let's say that the woman at point X is a light year away from the man at point Y. Einstein's theory of relativity forbids any kind of communication or information transfer between the two at any speed greater than the speed of light. Therefore, if she wants to send a message to him, it will take at least a year to get there. However, there appear to be experiences that happen more simultaneously than that. For example, if there is a particle at X and an entangled pair particle at Y, the measurement of some property at X would instantaneously establish the corresponding property of the particle at Y. How could this be possible without violating relativity?

At a more macroscopic level, many have experienced paranormal effects that defy the cause-and-effect logic of Newtonian mechanics, which seems to drive the rules of our apparent physical reality level; for instance, telepathy. Since there is no known mechanism for the transfer of information between minds at the apparent physical reality level, perhaps that communication occurs via the implicate order...

Such is the theory of those proposing a holographic paradigm to reality, one in which everything is interconnected at a deeper level of reality. Therefore, all information about everything exists in all places at any time, owing to the mechanism that allows access to this information in the implicate order level of reality.

In the case of a computer game (such as an MMORPG, a massively multiplayer online role playing game), the same model as above applies perfectly, except that we know that the apparent reality is really virtual, and the underlying mechanism that connects everything together is the computers and the network:

Everything Explained

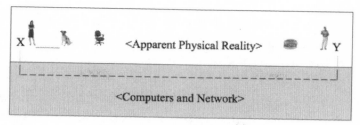

Figure 8.2

The woman and man are actually avatars that represent the corresponding people who are playing in the game. The real player could be anywhere—Ohio, Siberia, on a cruise ship. It doesn't matter as long as there is a fast enough connection to the network. Now, imagine that X and Y are really only separated by 100 meters in virtual space (in other words, the distances in the game that appear to be 100 meters, given the sizes of the objects and landscape rendered by the system), and the woman decides to throw an apple (not to scale) at the man.

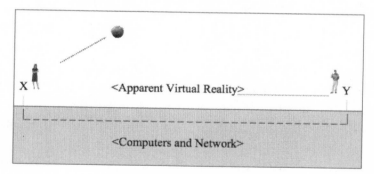

Figure 8.3

The rendering engine of the computer would calculate the apparent velocity of the apple and its trajectory from the perspective of each of the players, so it would appear to follow the laws of physics; that is, if the physics engine in the computer game were programmed well.

Let's say instead that X and Y were separated by light years, in galaxies far, far away from each other. If we were following the strict rules of the underlying physics engine, the man and the woman would not be able to communicate, or sense each other's presence or activity in any way. There would really not be much point creating scenarios in a video game like this if people can't interact in some way. But, of course, the game is not beholden to just the physics engine. In fact, there should be parameterized physics engines for different planets, which have different gravities and revolve around their stars at different rates.

But let's say we define a concept, like "the force" or a "teleporter" or a "super communicator" that bypasses normal physics and allows instantaneous communication. Now, the woman can just pull out her super communicator and tell the man why she is angry with him. Or she could teleport an apple to him mid-flight. Or simply use the force on him. Why is this possible? Because every few clock cycles of the computer, a new state of the entire game is calculated and updated. So, the player's attributes, objects in their environment and all influences on the player (such as sounds from the communicator or a teleported apple suddenly appearing out of nowhere and flying at his head) are updated effectively simultaneously regardless of where in the virtual universe they are. I say "effectively," because even the most sophisticated computers update data elements one at a time. (Note: some computers have multiple central processing units (CPUs), which could technically allow parallel processing. But even with these, the processes are usually split across the CPUs by application. This means that the entire game would still be running on only one CPU.)

A game or simulation has something called a frame rate, similar to the frame rate in movies or the refresh rate of a TV. It is the rate at which all elements of the game are updated. So, for example, if it is 60 fps (frames per second), then every 16.7 milliseconds there is a chance to make updates to every aspect of the simulation—the positions of the players, the positions of the objects in the game, the strength, health and magic attributes of the players, the elements in their inventory, any new communications between them, changes to their environment and any other changes to the virtual universe. Film makers have long since known that frame rates do not need to be that high for humans to experience smooth motion. The industry standard for many years was 24 frames per second. Our visual sensory apparatus perceives flickering around 30 fps or less, so movie projectors were designed to flash twice per frame. Therefore, although the "information frame rate" of the film was 24 fps, it was presented to us at 48 fps. Even high-definition formats, like IMAX, rarely use a frame rate above 48 fps.

The point to this discussion is that since all evidence points to our reality being digital and virtual, the simulation model matches very well and is extremely likely to be a valid model of reality. In such cases, our reality probably has a very high frame rate and follows a set of rules according to a physics engine, including a maximum speed limit of the speed of light. However, following the same model, we can have information transfer at a higher rate than the speed of light and, in fact, there would be nothing precluding some transfers or disturbances occurring essentially simultaneously. The holographic theorists ascribe the mechanics of this to the implicate order. To me, this is just a metaphor. The real mechanics are computational and ATTI/RLL is the computational mechanism.

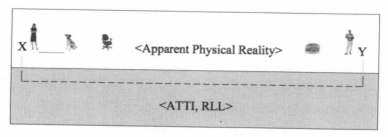

Figure 8.4

Matter and Forces—Explained

As discussed in Chapter 5, the evidence is strong that matter is simply data or information. Being digital, this information would be most efficiently represented in a binary form. So every physical object, from a quark to a giant sequoia, is ultimately represented by 1s and 0s, as shown below:

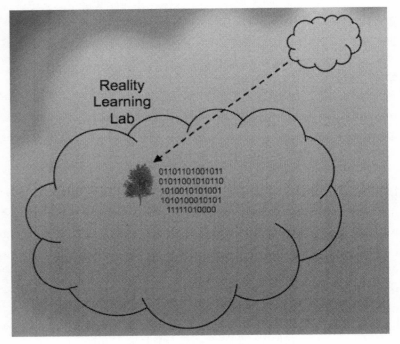

Figure 8.5

As described in Chapter 5, the RLL does not need to exhaustively encode all of the information contained in every single subatomic particle in the tree. That would be unnecessary, because nobody will be inspecting the existence and attributes of each particle. In fact, if you consider the amount of information needed to encode a giant sequoia fully, it might be around 1E34 bytes (3E31 atoms in the tree, a couple kilobytes to encode each position and assuming the most basic definition of a subatomic particle to be a single unique byte for each of the 200+ known types). However, encoding a tree to have the resolution and detail that is needed for conscious observers to experience it fully in our reality may not take more than a few megabytes. So the dynamic compression efficiency could theoretically be on the order of a very impressive 1E26 (1E34/1E8) and still meet the requirements of observational experience.

But it could be drastically different than that, as well. Perhaps the tree has much more information in it, so that RLL doesn't have to generate too much of it dynamically when a conscious observer digs into it and inspects it at a microscopic level. One thing appears certain, however—the compression ratio is greater than 1. If it weren't, there would be no such thing as the observer effect and entanglement.

So what are forces, then?

They are simply rules of the program; in gaming parlance, the "physics engine."

The four known forces of nature are the electromagnetic force, gravity, weak nuclear force and strong nuclear force. All of these forces are described classically by the concept of a field. Remember our discussion about models from Chapter 3? Fields are models of the idea that there are numbers (the strength of the field) associated with every point in space surrounding the object, which is responsible for the force. An example of a field is wind velocity. At each point in the space of the atmosphere, the wind is blowing at some speed and in some direction. The combination of a number (for example, wind speed) and a direction is called a vector. A vector in three-dimensional space (like the kind we think we live in) can be described by four numbers: 3 for the direction (how much it is pointing up, how much it is pointing east/west, and how much it is pointing north/south) and 1 for the magnitude (how strong it is). What does that sound like? A set of numbers for every point in space is nothing more than data.

In the case of our forces, the data describes how something should behave when it enters those points in space. In VR parlance, the data describes how something should behave when it enters those coordinates in the virtual space of the simulation. In the case of gravity, for example, an object at a certain point will be influenced to move in the direction and with the strength described by those numbers associated with that point. Now, the field vector can be calculated on the fly, which would probably be much more efficient than assigning sets of numbers to every point in space, for every object in space that can influence the force at that point. That is probably what is really going on in the RLL. But, the field idea is a convenient model for thinking about it and it works perfectly well for describing how objects behave near other objects. With the electromagnetic force, the idea of the field works the same way, except that there is another attribute to be considered in the objects under influence: charge.

In this way, all of the forces that we think we experience are nothing more than rules about how objects interact. Rules that can be easily calculated in a sufficiently powerful system, like RLL.

The Observer Effect — Explained

As will be shown in this chapter, Digital Consciousness Theory is an incredibly successful concept in terms of explaining the paradoxes and anomalies of quantum mechanics, including non-reality, non-locality, the observer effect, entanglement and even the retrocausality of John Wheeler's Delayed Choice Quantum Eraser experiment.

I came up with these explanations by thinking about how DCT could explain such curiosities.

But I thought it might be interesting to view the problem in the reverse manner. If one were to design a universe-simulating program, what kinds of curiosities might result from an efficient design? As a corollary, if a universe-simulation (such as RLL) were to evolve itself, what kinds of curiosities might result from an efficient design? For instance, data would probably be modeled in the following manner:

For any space unobserved by a conscious entity, there is no sense in creating the reality for that space in advance. It would unnecessarily consume too many resources.

For example, consider the cup of coffee on your desk. Is it really necessary to model every single subatomic particle in the cup of coffee in order to interact with it in the way that we do? Of course not. The total amount of information contained in that cup of coffee necessary to stimulate our senses in the way that it does (generate the smell that it does; taste the way it does; feel the way it does as we drink it; swish around in the cup the way that it does; have the little nuances, like tiny bubbles, that make it look real; have the properties of cooling at the right rate to make sense, etc.) might be 10 MB or so. Yet, the total potential information content in a cup of coffee is 100,000,000,000 MB, so there is a ratio of perhaps 100 trillion in compression that can be applied to an ordinary object.

But once you decide to isolate an atom in that cup of coffee and observe it (as noted in Chapter 6, it doesn't matter whether the observation is via an IC or simply the need to measure and record attributes of the atom), the program would then have to establish a definitive position for that atom, effectively resulting in the collapse of the wave function or decoherence. Moreover, the complete behavior of the atom, at that point, might be forever under control of the program. After all, why delete the model once observed, in the event (probably fairly likely) that it will be observed again at some point in the future. Thus, the atom would have to be described by a finite state machine.

What is a finite state machine (FSM)?

Finite State Machines

An FSM is simply a computational system that is identified by a finite set of states whereby the rules that determine the next state are a function of the current state and one or more input events. The FSM may also generate a number of "outputs," which are also logical functions of the current state.

The following is an abstract example of a FSM:

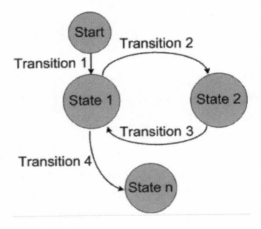

Figure 8.6

A computational system, like that laptop on your desk that the cat sits on, is by itself a finite state machine. Each clock cycle gives the system a chance to compute a new state, which is defined by a logical combination of the current state and all of the input changes. A video game, a flight simulator and a trading system all work in the same way. In 2016, the state changes in a typical laptop about 4 billion times per second. It may actually take many of these 250 picosecond clock cycles to result in an observable difference in the output of the program, such as the movement of your avatar on the screen. Within the big, complex laptop FSMs are many others running, such as each of those dozens or hundreds of processes that you see running when you click on your "activity monitor." And within each of those FSMs are many others, such as the method (or "subprogram") that is invoked when it is necessary to generate the appearance of a new object on the screen.

And so it appears to be with ATTI. If ATTI is finite, as was argued in Chapter 5, then it, too, is an FSM, albeit an incredibly complex one. And it will have FSMs modeling the behavior of all of its components, such as RLL. And RLL will have FSMs modeling the behavior of all of its components, such as trees and experimental apparatus for double-slit experiments. And such components will have FSMs modeling the behavior of the smallest subcomponent that can be observed. Because modeling anything deeper than that would be a waste of computational cycles.

In our thought experiment about isolating an atom in a cup of coffee through observation, we have had to create a little mini FSM to model the atom that has been isolated. In a similar way, when an IC observes the outcome of the double-slit experiment, the position, or *state*, of each particle flying through one of the slits is determined and stored into the local RLL memory system as shown in the following:

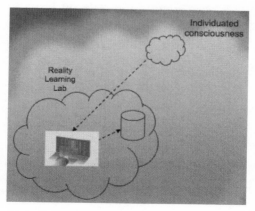

Figure 8.7

The behavior of the atomic FSM would be decided by randomly picking initial values of the parameters that drive that behavior, such as atomic decay. Thereafter, the FSM will run on its own, forever.

So, in summary, the process of "zooming in" on reality in RLL would have to result in *exactly* the type of behavior observed by quantum physicists. In other words, in order to be efficient, resource-wise, RLL decoheres only the space and matter that it needs to.

Recall that the purpose of ATTI is to grow and evolve, which means it has to learn. If you think about it, you realize that learning can only occur through free will; otherwise, all outcomes will be predetermined and no true learning can take place. And, in order to have free will, there must be probabilities to the states of the various elements of ATTI. If everything is deterministic, free will can't influence it. Therefore, observation *must* collapse the probabilistic states into definitive ones.

Wheeler's Single Electron—Explained

There is a concept in computer science called an "instance," which is similar to the idea of a template. As an analogy, consider the automobile. Every Honda that rolls off the assembly line is different, even if it is the same model with the same color and same set of options. The reason it is different from another with the exact same specifications is that there are microscopic differences in every part that goes into each car. In fact, there are differences in the way that every part is connected between two cars of equal specifications. However, imagine if every car were exactly the same, down to the molecule, atom, particle, string or what have you. Then we could say that each car is an instance of its template.

This would also be the case in a computer-based virtual reality. Every similar car generated in the computer program is an instance of the computer model of that car which, by the way, is an FSM. Each instance can be given different attributes, however, such as color, loudness or power. In some cases, such as a virtual racing game, where the idea of a car is central to the game, each car may be rather unique in the way that it behaves or responds to the inputs from the controller, so there may be many different FSMs for these different types of cars. However, for any program, there will be FSMs that are so fundamental that there only needs to be one of that type of object; for example, a leaf.

In RLL, there are also FSMs that are so fundamental that there only needs to be one FSM for that type of object. And every object of that type is merely an instance of that FSM.

In 1940, esteemed physicist and Nobel Laureate Richard Feynman received a phone call from esteemed physicist (and much deserved of the Nobel Prize) John Wheeler. As Feynman recalls: "I received a telephone call one day at the graduate college at Princeton from Professor Wheeler, in which he said, 'Feynman, I know why all electrons have the same charge and the same mass.' 'Why?' 'Because, they are all the same electron!'"[1] Although Wheeler's great insight was based on the concept of world lines through four-dimensional space-time, he may have been closer to the truth than anyone realized.

An electron is fundamental. It is a perfect example of an object that should be modeled by an FSM. There is no reason for any two electrons to have different rules of behavior. They may have different starting conditions and different influences throughout their lifetime, but they would react to those conditions and influences with exactly the same rules. Digital Consciousness Theory provides the perfect explanation for this. Electrons are simply *instances* of the electron FSM. There is only one FSM for the electron, just as Wheeler suspected. But there are many instances of it. Each RLL clock cycle will result in the update of the state of each electron instance in our apparent physical reality.

One electron FSM, many instances.

Quantum Entanglement — Explained

Armed with our model of reality and the concept of an FSM, entanglement suddenly loses its mystery, because particles are nothing more than FSMs. A particle is simply a set of information that describes the attributes of a particle (spin, position, velocity, charge, atomic mass), which can easily be encoded into a data structure. The dynamic behavior of the particle can be described by its FSM: event A acting on the particle in state X will cause the particle to move to state Y and possibly change the value of one or more of its attributes. State X could be the "undefined" state and event A is the process of observing the particle's position. So state Y would then be the defined state, with the position attribute populated by the calculated data element.

Let's play out the physics experiment that is so baffling. We will do it in two possible ways. In one case, we will assume that there is an FSM for each particle. In the second case we will assume that the same FSM is used for both...

Our physicist creates two particles in a lab from a known nuclear interaction and the system generates an FSM for each particle, because now we have created them. But since we don't really need to know their attributes, the FSM's data structures are left undefined and the FSMs are in state X. From a quantum mechanics point of view, an unobserved particle is a probability wave function, which means that its attributes, in particular *position*, are described probabilistically, instead of by a definitive number.

Interestingly, in computer systems, when some data is created that is inherently probabilistic, the process of creating it involves something called a random number seed. This is a number that is given to a process that generates the random data. Effectively, it isn't truly random, because if you always give the process the same seed, it will always generate the same sequence of data points. That is why it is call pseudo-random. Computers typically use seeds that are almost guaranteed to be different each time the process is invoked, for example, by using the clock as input to the seed.

A potentially significant side effect of this process is that if two FSMs were created at the same time using the same random number seed, they would behave, in some sense, in lockstep. This feels similar to what happens with entangled particles—once created, they continue to behave in lockstep. In any case, one of the attributes of each particle, stored in the FSM, would be the link to the paired or *entangled* particle. The data structures of each FSM could look something like this (note: the details may be filled in by any particle physicist):

State
Position
Link to entangled pair
Charge
Spin

Next, the particles are separated by an arbitrary distance, but still not measured. No need to change anything in the FSMs.

Now, the experimenter makes a measurement on one of the particles and the FSM's data structure is updated to include all of the attributes of that particle. Effectively, the particle has now *come into existence*. At this point in time, the entangled particle gets its attributes, because part of the program is to update the entangled particle(s) in the same cycle as the original.

Alternatively, it is possible that only a single data structure is created, because the particles are entangled by virtue of the fact that they were created at the same time. One attribute may be a linked list of references to the entangled particle(s). In this case, when the measurement is done, the programming is even simpler. Both of the particles' attributes get established at the same time, by virtue of the fact that they are using the same FSM. Two apparent particles in reality, but one underlying mechanism to describe both. In the case where each entangled particle must have a complementary property, such as spin, this is easily handled in the FSM. The data structure could be something like this:

State
Position A (x,y,z)
Position B (x,y,z)
Charge
Spin A

The FSM contains the position of both particles, the charge of both and the spin of one of them. The complementary spin will always be rendered in the other by the FSM.

Both of these models provide equally valid explanations for the behavior of entangled particles:

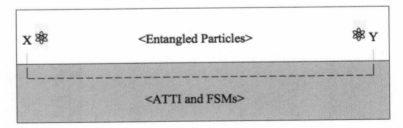

Figure 8.8

After the measurement event, at each cycle through the main program loop, whatever one particle does, its synchronized counterparts also do the same. Since the program operates outside of the virtual laws of physics, those particles can be placed anywhere in the program's reality space and they will always stay synchronized. Yet, their motion and other interactions may be subject to the usual physics engine.

The Quantum Zeno Effect—Explained

Lurking amidst the mass chaos of information that exists in our reality is a little gem of a concept called the quantum Zeno effect. It is partially named after ancient Greek philosopher Zeno of Elea, who dreamed up a number of paradoxes about the fluidity of motion and change. For example, the "Arrow Paradox" explores the idea that if you break down time into "instants" of zero duration, motion cannot be observed. Thus, since time is composed of a set of instants, motion doesn't truly exist. We might consider Zeno to have been far ahead of his time, as he appeared to be thinking about discrete systems and challenging the continuity of space and time a couple thousand years before Alan Turing resurrected the idea in relation to quantum mechanics:

> It is easy to show using standard theory that if a system starts in an eigenstate of some observable, and measurements are made of that observable N times a second, then, even if the state is not a stationary one, the probability that the system will be in the same state after, say, one second, tends to one as N tends to infinity; that is, that continual observations will prevent motion ...[2]

The term "quantum Zeno effect" was first used by physicists George Sudarshan and Baidyanath Misra in 1977 to describe just such a system—one that does not change state because it is continuously observed.

The challenge with this theory has been in devising experiments that can verify or falsify it. However, technology has caught up with philosophy and, over the last 25 years, a number of experiments have been performed, which seem to validate the effect. In 2001, for example, physicist Mark Raizen and a team at the University of Texas, Austin, showed that the effect is indeed real and the transition of states in a system can be either slowed down or sped up simply by taking measurements of the system.[3]

The quantum Zeno effect might not be observed in every case. It only works for non-memoryless processes. Exponential decay, for instance, is an example of a memoryless system. Frequent observation of a particle undergoing radioactive decay would not affect the result, because an exponential probability is always the same at each point in time.

[As an aside, I find it very interesting that a "memoryless system" invokes the idea of a programmatic construct. Perhaps with good reason...]

However, in theory, a system with memory, or "state," is subject to the quantum Zeno effect. It will manifest itself by appearing to reset the experiment clock every time an observation is made of the state of the system. The system under test will have a characteristic set of changes that vary over time. In the case of the University of Texas experiment, trapped ions tended to remain in their initial state for a brief interval or so before beginning to change state via quantum tunneling, according to some probability function. For the sake of developing a clear illustration, let's imagine a process whereby a particle remains in its initial quantum state (let's call it State A) for 2 seconds before probabilistically decaying to its final state (B) according to a linear function over the next second. Figure 8.9 shows the probability of finding the particle in State A as a function of time. For the first 2 seconds, of course, it has a 0 percent probability of changing state, and between 2 and 3 seconds it has an equal probability of moving to State B at any point in time. A system with this behavior, left on its own and measured at any point after 3 seconds, will be in State B:

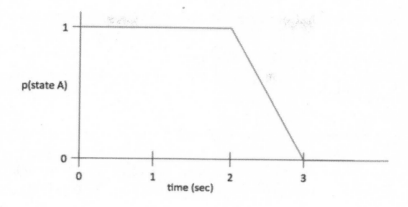

Figure 8.9

What happens, however, when you make a measurement of that system, to check to see if it changed state, at t=1 second? Per the quantum Zeno effect, the experiment clock will effectively be reset and now the system will stay in State A from t=1 to t=3 and then move to State B at some point between t=3 and t=4. If you make another measurement of the system at t=1, the clock will again reset, delaying the behavior by another second. In fact, if you continue to measure the state of the system every second, it will never change state. Note that this has absolutely nothing to do with the physical impact of the measurement itself; a 100 percent non-intrusive observation will have exactly the same result.

Also note that it isn't that the clock doesn't reset for a memoryless system, but rather that it doesn't matter, because you cannot observe any difference. One may argue that if you make observations at the Planck frequency (one per jiffy), even a memoryless system might never change state. This actually approaches the true nature of Zeno's arguments, but that is a topic for another time; it is a concept that is much more philosophical than falsifiable. In fact, "quantum Zeno effect" is a misnomer. The non-memoryless system described above really has little to do with the ad infinitum inspection of Zeno's paradoxes, but we are stuck with the name. And I digress.

Anyway, why would this happen?

It appears to be related in some way to the observer effect and to entanglement:

- Observer effect—Once observed, the state of a system changes.
- Entanglement—Once observed, the states of multiple particles (or, rather, the state of a system of multiple particles) are forever connected.
- Quantum Zeno—Once observed, the state of a system is reset.

What is common to all three of these apparent quantum anomalies is the coupling of the act of observation with the concept of a state. For the purposes of this discussion, it will also be useful to invoke the computational concept of our friend, the FSM.

I have explained the observer effect and entanglement as logical necessities of an efficient programmed reality system. What about quantum Zeno? Why would it not be just as efficient to start the clock on a process and let it run, independent of observation? A clue to the answer is that the act of observation appears to create something.

In the observer effect, it creates the collapse of the probability wave functions and the establishment of definitive properties of certain aspects of the system under observation (for example, position). As explained, this is not so much a matter of efficiency as it is of necessity, because without probability, free will doesn't exist and without free will, we can't learn. If the purpose of our system is to grow and evolve then, by necessity, observation must collapse probability.

In entanglement, the act of observation may create the instantiation of an FSM, which subsequently determines the behavior of the particles under test. Those particles are just data, as I have shown, and the data elements are part of the same variable space of the state machine. They both get updated simultaneously, regardless of the "virtual" distance between them.

In quantum Zeno, the system under test is in probability space. The act of observation "collapses" this initial probability function and kicks off the mathematical process by which future states are determined based on the programmed probability function. But that is now a second level of probability function; call it probability function 2. Observing this system a second time now must collapse the probability wave function 2. But to do so means that the system would now have to calculate a modified probability function 3 going forward—one that takes into account the fact that some aspect of the state machine has already been determined (for example, the system has or hasn't started its decay). For non-memoryless systems, this could be an arbitrarily complex function, since it may take a different shape for every time at which the observation occurs. A third measurement complicates the function even further, because even more states are ruled out.

On the other hand, it would be more efficient to simply reset the probability function each time an observation is made, owing to the efficiency of the reality system.

The only drawback to this algorithm is the fact that smart scientists are starting to notice these little anomalies; although the assumption here is that the reality system "cares." It may not. Or perhaps that is why most natural processes are exponential or memoryless—it is a further efficiency of the system. Man-made experiments, however, don't follow the natural process and may be designed to be arbitrarily complex, which ironically serves to give us this tiny little glimpse into the true nature of reality.

What we are doing here is inferring deep truths about our reality that are in fundamental conflict with the standard materialist view. This will be happening more and more as time goes forward, and physicists and philosophers will soon have no choice but to consider Digital Consciousness Theory as their ToE.

Delayed-choice Quantum Eraser—Explained

Seemingly in an effort to push the boundaries of quantum strangeness, in 1978 John Wheeler dreamed up the mother of all quantum *gedankenexperimenten* (thought experiments)—the *Delayed Choice Experiment*.[4] Recall the setup of the double-slit experiment. If a detector is placed in such a way as to determine which slit a photon would go through, the photon will act as a particle and be detected. If no detector is there, it will appear to go through both slits simultaneously as a wave. So the decision to detect impacts the outcome of the experiment.

What if, Wheeler proposed, the decision to detect occurs *after* the particle has already gone through the detectors one way or another? It would appear that the "decision" of the particle to act as a particle or a wave would have to happen at measurement time, thereby inferring its path retroactively. Such logic could be applied to an extreme case of a photon traveling for millions of miles and bending around an intervening galaxy in one direction or the other. We can determine which way the photon traveled around the galaxy by measuring its direction. Or we can decide *not* to make that measurement and observe the interference pattern it would make because, as a wave, it went both directions simultaneously. So how could the decision to detect a direction today influence the path that the photon took millions of years ago? It seems to defy logic. But that is only because what we call logic includes the concept of causality.

In any case, the technology to implement such an experiment was beyond the technology of Wheeler's time in 1978, but it isn't any more. In 2000, Yoon-Ho Kim, R. Yu, S. P. Kulik, Y. H. Shih and Marlan O. Scully performed this remarkable experiment using the setup shown in Figure 8.10:

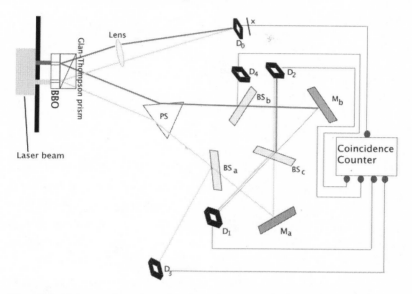

Figure 8.10 (reproduced under the GNU Free Documentation License from the Wikimedia Commons)[5]

The experiment also contains the twist of "quantum erasing," whereby the photons are "marked" in some way (for example, polarization). The marking results in the disappearance of the interference pattern, because now the paths are known. But when the mark is effectively erased by a combining and repolarizing, the interference pattern comes back, because it is now not known which slit those particular photons went through.[6]

Experiments such as the one performed by Kim et al. are so astounding in their implications that scientists rigorously analyze them for loopholes, such as esoteric explanations that fit within the realm of experimental probability and exploit some unlikely nuance of traditional physics. In this case, it was felt that because all of the experimental apparatus was together in the same room, it was impossible to rule out some sort of causal slower-than-light communication between the choice of measurement elements and the observation elements in the setup.

A subsequent delayed-choice quantum eraser experiment performed at the University of Vienna in 2013, however, ruled out that possibility by separating the elements of the experiment by 144km between two of the Canary Islands.[7]

And in 2015, physicists at the Australian National University (ANU) performed the delayed-choice experiment using atoms rather than photons with the same result. "At the quantum level, reality does not exist if you are not looking at it," said Associate Professor Andrew Truscott. "The atoms did not travel from A to B. It was only when they were measured at the end of the journey that their wave-like or particle-like behavior was brought into existence."[8] Most scientists argue that this is not really a violation of causality — it merely appears so. Instead, it is a violation of objective reality. The photon or atom did all things in a probability space and we don't actually determine reality until we make a measurement.

I'm not sure which is the more astounding result!

Digital Consciousness theory explains it all perfectly well. The portion of the RLL FSM that controls the behavior of the atoms or photons in the experiment doesn't really come into existence until the measurement is made, which effectively means that the atom or photon doesn't really exist until it needs to. In RLL, the portion of the system that needs to describe the operation of the laser, the prisms and the mirrors, at least from the perspective of the observer, is defined and running, but only at a macroscopic level. It only needs to show the observer the things that are consistent with the expected performance of those components and the RLL laws of physics.

So, for example, we can see the laser beam. But only when we need to determine something at a deeper level, like the path of a particular photon, is an FSM for that proton instantiated.

And in the delayed-choice quantum eraser experiment, that FSM only starts when the observation is made, which is *after* the photon has gone through the apparatus; hence, it never really had a path. It didn't need to. The path can be *inferred* later by measurement, but it is incorrect to think that that inference was objective reality. There was no path.

"There is no spoon."
- *Neo, in* The Matrix

There are only the attributes of the photon determined at measurement time, when its FSM comes into existence. Again, the photon is just data, described by the attributes of the FSM, so this makes complete sense. Programmatically, the FSM did not exist before the IC required a measurement, because it didn't need to. Therefore, the inference of "which path" is *only* that—an inference, not a true history (see Figure 8.11).

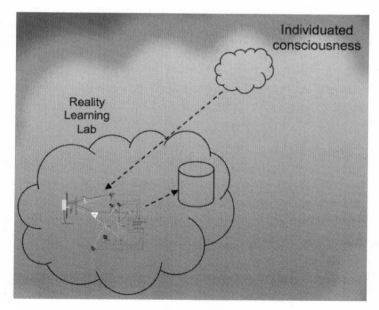

Figure 8.11

The Finely Tuned Universe—Explained

Recall the dazzling set of attributes of our reality that seem to have been perfectly selected, aka "the Goldilocks enigma," already presented in Chapter 6, such as the fact that universal constants cancel out all of the vacuum energy to an amazing accuracy of 1 part in 10^{115}. We presented a logical argument that such a set of apparently anomalous coincidences points more likely to a reality that was designed rather than one that sprung from nowhere, as the Many Worlds/Anthropic Principle apologists would have you believe.

But there is another theoretical alternative to the anthropic principle that might keep the solution to this embarrassing scientific anomaly somewhat in the realm of science. It is called Flexi-laws. Originally proposed by physicist John Wheeler, the concept argues that physical laws are not immutable and may actually have been evolving ever since the Big Bang. Couple that with our old quantum mechanical friend, the Heisenberg Uncertainty Principle, and you get Stephen Hawking's quantum cosmology, which asserts that the past (and therefore the evolution of our physical laws) is influenced by the present state of the universe. Wheeler suggests that "the existence of life and observers in the universe today can help bring about the very circumstances needed for life to emerge by reaching back to the past through acts of quantum observation."[9]

Doesn't Occam's razor apply here, also? Are we to believe that through 13 billion years of quantum uncertainty we are able to influence the past in such a way as to reverse-evolve existing laws of physics and chemistry via the Flexi-laws concept? Or, again, would it be simpler for the universe to come about via a system and a fundamental evolutionary principle?

There is an important nuance (and philosophical question) here. Was the universe designed for matter and life by some intelligent entity, or did it evolve to become designed for matter and life? The distinction might seem to be arbitrary, because it is only the end result that matters. Yet, it might be interesting to take a stab at the compilation of evidence for either of these two scenarios.

There isn't really any evidence for a designer per se. People who have had mystical experiences, OBEs, NDEs and past-life recollections don't tend to speak of encountering a single entity. They do tend to experience a feeling of love and bliss, however, but it typically emanates from a global consciousness in which we are all connected, rather than a separate entity. There is also a universal pattern of evolutionary progression that tends to exist at many scales or levels. This pattern is that possibilities that generate productive change tend to be reinforced, whereas possibilities that generate negative change tend to be diminished.

This can be seen in mutations on the microevolutionary scale and is responsible for species evolution. It can be seen on the human level, as our society tends to evolve toward more productive and profitable patterns. It is easy to complain about society, and there are certainly negative and evil patterns throughout the world, but when one considers things like murder rates, starvation rates, disease, humans' treatment of other humans and other animals, it is easy to see that we do progress.

Societies that develop negative patterns (such as totalitarianism) tend to fail. Societies that embody healthy patterns (such as egalitarianism) tend to flourish. And we also see patterns of developing complexity on a galactic level, such as the evolution from a featureless gas state of the universe, to one with clumpy galaxies and stars that support complex elements and life systems. So, for a number of reasons, I believe that our universe (in this context, the apparent physical universe) has evolved on its own.

What is it? It is simply RLL, the Reality Learning Lab. And it evolved according to a system similar to the model in Figure 8.12:

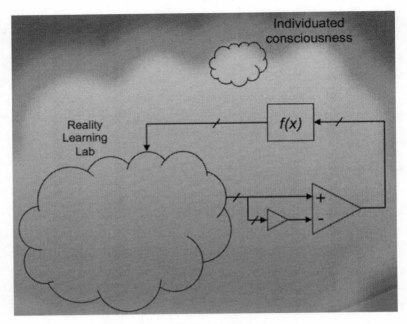

Figure 8.12

where...

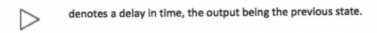
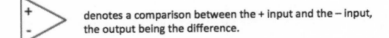

Figure 8.13

The Global Consciousness System or ATTI evolved the RLL in order to give us ICs a construct in which to learn and evolve. But it did it, according to physicist Tom Campbell, by following the fundamental law of continuous improvement, as explained in Chapter 7. This law works in the following manner: some aspects of the RLL are measured. In the very early epoch of this system, it would have included a measure of the ability of FSMs to combine in an intelligent manner to form more complex machines. Later, it probably included a measure of the correlation of the ability of each IC to learn from their experiences with the rate of forgetting past experiences in the RLL.

Each set of data is compared to the previous output metrics of the system. If the result is favorable, the system is either left alone or enhanced in a manner that is in line with the previous adjustment. If the result is unfavorable, the system either goes back to the previous state or is modified in a manner that opposes the previous adjustment. In this way, the system can continuously evolve to be more ideal. The parameters of the physics of the system will of course end up being set to be ideal for the apparent formation of matter, planets and life, not because they were selected to be such by a separate conscious entity, but because they were selected to be such and/or evolved to be such by ATTI.

Meditation and Intuition—Explained

"If you are depressed, you are living in the past.
If you are anxious, you are living in the future.
If you are at peace, you are living in the present."
- Lao Tzu

The blog Mental Health Daily lists dozens of National Institute of Health studies and research papers providing evidence for an amazing number of mental and physical health benefits of meditation. These include reduced blood pressure, increased brain matter, improved IQ, memory and cognitive abilities, reduced stress and anxiety, reduced depression, pain management, increased creativity, enhanced immune system, enhanced spiritual perception and general wellbeing.[10]

Wow! That's a very impressive list. How can a single practice have such positive impacts on so many seemingly disparate attributes? The answer is actually pretty simple once you realize who you are and what meditation is. You are digital consciousness, essentially consisting of information that describes every single aspect of your being. Meditation is simply a practice that facilitates the modification of that data, thereby modifying those attributes. Here's how it works (see Figure 8.14)...

We start with our usual suspects—the RLL and our IC, as shown above. I've taken the liberty of not making the graphic to scale, so that we can jam into the IC all sorts of stuff. IC can be thought of as being the same as "mind." We are our mind, no more, no less.

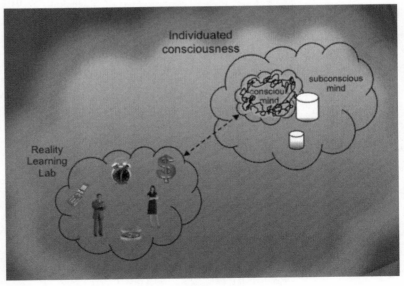

Figure 8.14

But there are two kinds of mind—conscious and subconscious—

as shown in the figure. The conscious mind is what is dealing with your current awareness—sensations, perceptions, some memories, feelings and fantasies. The subconscious mind, on the other hand, is always working in the background, processing everything that occurs in the periphery of our senses, such as subliminal perceptions and sounds that don't register. It is handling autonomous functions like breathing, blinking and swallowing, and it stores all of the complex information that allows us to retain skills we may have learned decades ago, like riding bikes, driving cars and dealing with social situations. It is constantly making associations between present perceptions and memories, and it pulls out all of that information that you need to handle every single daily task. Implicit thought, implicit memory and implicit learning goes on without you even realizing it, all thanks to the subconscious mind.

Research has shown that the conscious aspect of your mind is actually very slow and deals with very little information. By some estimates, the conscious mind processes between 16 and 40 bits of information per second,[11, 12] and can only keep track of 4–7 items at any one time.[13]

By contrast, the subconscious mind processes between 20 Mbits[14] and 400 Gbits[15] per second, and stores around 2.5 petabytes of information;[16] that is, only about a billion times faster than the conscious mind, with a quadrillion times more storage capacity. This is why, in the figure above, I show the temporary life data as being mostly in the subconscious, but a small amount can also be in the conscious mind.

So it is the subconscious mind that is responsible for all of the things that make us who we are—our self-image, memories, immune system, values, traits and the complex process for the myriad decisions that we make every instant. This is the place to go to if you want to reduce your blood pressure, increase your brain matter, improve IQ, memory and cognitive abilities, reduce stress and anxiety, reduce depression, manage pain, increase creativity, enhance the immune system, enhance spiritual perception and improve your general wellbeing. Sound familiar?

We have direct access to all of that information that makes us who we are. So why can't we change ourselves at will? And make ourselves superhuman, gorgeous, perfect and with the ability to live forever. The answer is because those things are too complicated to adjust to perfection. Humans are complex adaptive systems. We can't be analyzed and tweaked to perfection, because there are too many moving parts.

Take as an analogy, a city. It is dirty, jammed with people and cars, and some of the people are down on their luck. How do you attempt to make the city perfect? Let's tackle the traffic problem by widening all of the streets so that there is less traffic. That would entail tearing down buildings, disrupting businesses and putting people out of work, which would result in more crime and more homelessness. Every problem that you attempt to fix will have feedback effects and ramifications on many other problems.

Complex adaptive systems include ecosystems, traffic, businesses, social interactions and yes, every single living entity on the planet. The field of complex adaptive systems has been well researched and developed. Without writing another book in this chapter, I'll summarize by saying that the way to improve complex adaptive systems is not through rigorous analysis and quick fixes. It is through probing, sensing and responding. Try something to see what happens. Let the system settle. It is better than before? No? Undo what you just did. Yes? Keep doing that and now try something to further improve the system. Evolution works this way. And so do we.

Our IC is a highly complex adaptive system. Our purpose in each life is to improve the quality of that system—aka our soul. That can only be done slowly, subtly, through experimentation, feedback and continuous improvement. So, the system has evolved only to allow us to make subtle changes and influences. Otherwise, the response of our system would be chaotic.

Therefore, yes, we *can* influence all of those things that meditation has been shown to be good for, but we can only do it slowly, weakly and subtly. But to be able to do it at all, our mind must have access to the subconscious. And there's the rub. Because our conscious mind is chock full of the issues of the day.

As Figure 8.14 shows, our mind is cluttered up with money, bosses, chores to do, schedules to keep, what to have for dinner, what I should say to whomever just texted me and what happened last night on *Keeping Up with the Kardashians*. This stuff is necessary—well, some of it is—for us to get the learnings in this lifetime that we need. But it also gets in the way of the kinds of subtle improvements we can make to ourselves by accessing the subconscious mind.

And so, the value of meditation is to help us to clear out that junk that is clogging up the conscious mind, so that we can access the subconscious mind.

Merely accessing the subconscious with a little more clarity gives us a lot of things that we wouldn't otherwise have, such as access to deep memories and feelings, deeply stored knowledge and an awareness of the massive amount of subliminal data that we continuously collect. This is also known as intuition.

"The intellect has little to do on the road to discovery. There comes a leap in consciousness, call it intuition or what you will, and the solution comes to you, and you don't know how or why."
- *Albert Einstein*

"The only real valuable thing is intuition."
- *Albert Einstein*

"My [trading] decisions are really made using a combination

of theory and instinct. If you like, you may call it intuition."
- *George Soros*

But accessing the subconscious, perhaps most importantly, is also the gateway to the rest of ATTI. This latter point is also known as spirituality. This is why meditation is the best path to a spiritual awakening.

But what about those improvement modifications that we wish to make to ourselves via the subconscious? First, use meditation to clear the mind and then use other techniques, such as visualization, to effect the change (Figure 8.15).

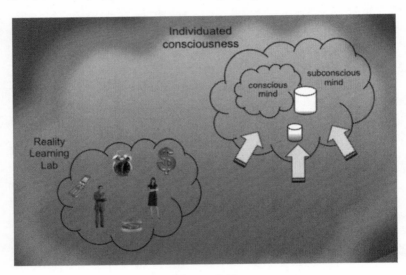

Figure 8.15

Nature vs. Nurture — Explained

Research supports the idea that "many aspects of the child's present personality have carried forward intact from the past life: behavior, emotions, phobias, talents, knowledge, the quality of relationships, and even physical symptoms."[17] Sadly, such research is as heretical to scientific orthodoxy as heliocentrism was 500 years ago; although referring to it as epigenetics may be a safe way for scientists to dip their toes into the water without getting scalded.

What is going on is shown below in Figure 8.16. When a new soul incarnates, it has associated with it some genetic data related to the biological parents of the new body. It also has associated with it a new database, blank at birth, of temporary life data, which accumulates as the person experiences physical reality. But, it also carries along with it all of the previously accumulated soul data that holds the learnings and values that we carry from life to life. This information represents all that has evolved about our IC.

The learnings that we attained in previous lives may very well be gender specific. It isn't hard to imagine that you learn different things in this world as a male than you do as a female, especially during previous epochs of human evolution. Hence, the affinity for a particular gender. This also perfectly explains the stark difference between value systems and personality traits of twin siblings. Their soul histories may have been completely

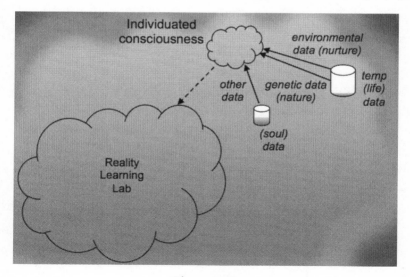

Figure 8.16

different. There is some evidence (re. Brian Weiss and others) that people tend to cluster together, selecting new lives that are more familiar than randomly different and including some of the same ICs in different lives (the true meaning of soul mate). However, in order to attain certain learnings in the new lifetime, one might be likely to select a drastically different community to be born into, which would account for the very differences that cannot be explained through nature or nurture.

So, the next time someone gets on your case for not living up to the family ideal, smile knowingly and be proud of your karmic heritage.

Paranormal — Explained

We showed in Chapter 6 how much strong statistical evidence there is for paranormal phenomena, such as telepathy and remote viewing. And yet these phenomena are unpredictable both in terms of occurrence and intensity. Some refer to them as subtle forces, but at times they are not at all subtle. What could possibly explain phenomena that can range from nearly imperceptible to intense, phenomena that only seem to happen to certain people but can occur with a wide range of likelihoods, phenomena that generally can't be controlled but often appear under highly charged conditions, or phenomena that can actually be more likely to occur with practice?

Imagine that you turn on your faucet and nice clean cold water comes out. Then you notice that it has a slight greenish hue to it. You pull out your handy-dandy water purity test kit and find that your water contains some copper oxide — say, 1 molecule of copper oxide for every million molecules of water. The copper oxide might be leaking into the water supply from an oxidized copper pipe somewhere.

The same can happen with data. Imagine that you have a network cable connected to the wall and you are streaming a movie at 4 Mbits/second (that's 4 million bits of data per second). Your communication channel is that network cable and your data is the stream of bits that encodes the movie that you are watching. Now, imagine that somewhere in the network, someone or something has injected a low bit rate stream of, say, National Security Agency (NSA) data, along with your movie, perhaps 100 bits/second. So, for every 40,000 bits of movie data, you are also receiving a bit of NSA data, encoding, say, an unauthorized phone recording of some random dude in Iowa talking to his drug dealer. Your movie player would completely ignore the excess bits and via the magic of error correction codes, would play the movie with no degradation or noise.

However, the data is there. You just aren't using the right application to extract it from the data stream. Effectively, your TV or computer is "focused" only on the movie and is ignoring the other data. But, if you had the right application that could extract the excess data from the data stream, decode it and play it on some other application, such as an audio player, you would then be aware of two streams of data. The NSA stream is so *tenuous* that it is barely there. But, with proper *focus*, you can detect it.

And so it is with our reality. In Figure 8.17 below, we see how various paranormal experiences work. Your IC is connected to the RLL, where it experiences a tree, for example. You are interacting with this *local* tree, which is described entirely by data. You perceive it by receiving data about its makeup—how it looks, how it feels. You send data to RLL whenever you do something that causes the tree to respond, such as pulling a leaf off its branches. Meanwhile, there is another tree somewhere far away (in our apparent physical reality) that other ICs interact with. This *remote* tree is also just composed of data—data that exists in the same RLL space, but because it is not in your physical proximity in your reality, RLL doesn't make it easy for you to experience it. If it were easy, we would be bombarded

223

with too much information to process, most of it irrelevant to the current experience. However, the data is there and your consciousness could theoretically access it. In fact, paranormal experiences show that it can be accessed.

Remote viewing, for example, can be easily explained by the leaky data model. The data stream that describes the remote tree can be thought of as leaky data that is getting into your main stream of conscious data. It can also be thought of as *out-of-band* data, as I discussed in *The Universe — Solved!* In information theory, out-of-band data is data that is in a separate communications channel. It is actually something that occurs every time you make a phone call. "In band" refers to the communications channel that you speak over; the pair of wires or the channel of bits that encodes your voice, which are sent as you are communicating with the person at the other end of the "line."

In the early days of telecommunications, the signaling of the number to be called was carried over the same pair of wires. As the network became digital, it was found to be far more scalable and flexible to send the signaling data (phone number or address of the person being called) over a completely separate network. This was referred to as "out-of-band" signaling.

In the movie *The Matrix*, the agents wore earpieces that told them what was going on at all times in the matrix. That communication was an out-of-band channel, in that it was not being sent through the standard reality that they were in. For another example, consider the MMORPG virtual reality game. Your character communicates with other characters by typing or speaking. The communication occurs in the game reality. There may be rules that allow you to speak directly to and only to another character or it may be that every character that wants to listen is capable of hearing any given communication.

Imagine that you want to speak to another character without being monitored by any other character. How could you do it? If you know the phone number of the person playing that character, you could simply call them up, send an email, IM, chat or use any other valid communication mode that is outside of the game world. What you have done is communicate "out of band."

Both out-of-band data and leaky data are just models for how paranormal data makes it into your consciousness. Through practice and the following of rigorous protocols, it is possible to tap into this leaky data stream. Not surprisingly, based on the research discussed in the prior section on meditation, quieting your conscious mind is core to the remote-viewing practice. Essentially, you are minimizing the flow of local data, so that the flow of leaky data is more noticeable. Let's say, for example, that the leaky data is 100 bits per second and that during a particularly immersive reality experience, your local data stream is 100,000 bits per second (aka 100 kbps). Clearly the 100 bps data, occupying 0.1 percent of your RLL communications channel, is effectively lost in the noise.

But what if, through meditation, you were able to quiet your mind and tune out 99 percent of your local traffic. Then you would still be receiving 10 times as much local sensory data as the remote sensory data that you asked for via the appropriate protocol. But that remote viewing data would now be noticeable, wouldn't it? It's still below the level of conscious experience, and might come to you in the form of vague shapes and patterns, exactly as was described by the "men who stared at goats."

In the case of telepathy, the stream of information doesn't even need to go through the RLL. It can be a direct connection with another IC in ATTI. Again, we don't normally perceive direct connections to other conscious entities with whom we are not interacting in the RLL. If that were common, we would be bombarded with too much information to process, most of it irrelevant to the current experience. However, with the proper combination of quieting the conscious mind, intent, focus, serendipity and other factors known only to ATTI, it could be possible to sense the thoughts of another conscious entity.

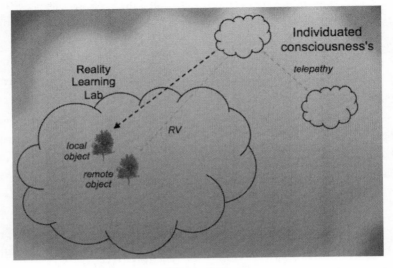

Figure 8.17

One might wonder what the purpose is for these leaky data streams. By our fundamental law of continuous improvement, we can safely assume that ATTI and the RLL always evolve to be increasingly efficient, useful and spiritually productive. Therefore, there must be some value to leaky data.

In some cases, it can be extremely valuable. There are instances, for example, where individuals have had strong premonitions about impending disasters in their lives and acted on those premonitions to save their lives. A case in point is Eva Hart, then a 7-year-old girl traveling across the ocean with her parents, on the RMS *Titanic*. Her mother, Esther, had a strong precognitive feeling about the impending disaster. The night before the accident she recalls being awakened "frozen with terror of I know not what. Then I stood up and shook my husband, who was still sleeping soundly. 'Ben,' I said, 'Ben, wake up, get up, something dreadful has happened or is going to happen.'" As Eva remembers:

This was the only time that she ever had a premonition and she was never able to think of any reason why she should have had such a strong pre-cognition just on that one occasion. There are many people who claim they have premonitions, but I have never met anyone who was so convinced as she was, to the extent that she was prepared to sit up every night and even to be ridiculed throughout the voyage.[18]

That premonition caused the entire family to act with a haste and lack of confusion when the crash occurred, and Eva remarked, "That we were saved was due solely to my mother's premonition."

There are also cases where people didn't pay close attention to their premonitions. People do in fact have free will to act or not act on this leaky data. According to the History Channel website, for example, "Ward Hill Lamon — Abraham Lincoln's former law partner, friend and sometime bodyguard — told a famous story about the 16th U.S. president's premonition of his own death. According to the tale, just a few days before his assassination on April 14, 1865, Lincoln shared a recent dream with a small group that included his wife, Mary Todd, and Lamon. In it, he walked into the East Room of the White House to find a covered corpse guarded by soldiers and surrounded by a crowd of mourners. When Lincoln asked one of the soldiers who had died, the soldier replied, 'The president. He was killed by an assassin.'"[19] On the day that John F. Kennedy was assassinated, he noted that "last night would have been a hell of a night to assassinate a president. Anyone perched above the crowd with a rifle could do it."[20] And, according to the book *The Day John Died*, by Christopher Andersen, "In the latter years of her life, Jackie [Onassis] had a recurring premonition that John would be killed piloting his own plane. She pleaded with Maurice [Tempelsman, her longtime companion] to do whatever it took to keep John from becoming a pilot."[21]

Another paranormal anomaly easily explained by the data stream concept is Michael Talbot's extraordinary hypnosis experience recounted in Chapter 6. Hypnosis can have a powerful effect on the mind, which resides in the IC outside of the RLL. In Talbot's example, subject Tom was convinced that his daughter, Laura, was not in the room. Her representation (aka avatar) was indeed part of the RLL representation of the room that the other conscious entities were interacting with. However, Tom, thinking that she was not there, broke with that portion of the illusion of consensus reality. Hence, the data stream between the RLL and his IC either did not contain Laura's data (but did contain the view of the back side of the watch) or it contained all of the data and he only processed it as if Laura's portion didn't exist. The distinction is insignificant and only gets to the actual mechanics of Digital Consciousness, which is most likely forever beyond our ability to determine.

Precognition—Explained

"I don't believe in yesterday."
- *John Lennon*

In addition to the many anecdotal cases of precognition, including those mentioned in the previous section, there has also been a great deal of scientific research that supports precognition as a bona fide experience. Dr. Daryl Bem, Professor Emeritus of Psychology at Cornell University, published an astounding paper in the *Journal of Personality and Social Psychology* in 2011 called "Feeling the Future: Experimental Evidence for Anomalous Retroactive Influences on Cognition and Affect."[22]

In plain English, he draws on the results of 8 years of scientific research to prove that precognition exists. His research techniques utilized proven scientific methods, such as double-blind studies. In each case, he reversed the sequence of well-studied psychological phenomena, so that "the event generally interpreted as the cause happened after the tested behaviour rather than before it."[23] Across all of the studies, the probability of these results occurring by chance and not owing to a real precognitive effect was calculated to be about 1 in 100 billion.

This little scientific tidbit went viral quickly, with the Twitterverse and Reddit communities posting and blogging prolifically about it. I have to commend the courage that Dr. Bem had in submitting such an article and that the APA (American Psychological Association) had in accepting it for publication. Tenures, grants and jobs have been lost for far less of an offense to the often closed-minded scientific/academic community.

More to the point, though, this had many scientists scratching their heads. What could it mean about our reality? As we have seen, quantum physicists say that objective reality doesn't really exist, anyway. But, most scientists from other fields have compartmentalized such ideas to a tiny corner of their awareness labeled "quantum effects that do not apply to the macroscopic world." Guess what? There isn't a line demarking quantum and macroscopic, so we need to face the facts. The world isn't as it seems and Daryl Bem's research is probably just the tip of the iceberg.

OK, what could explain this?

Conventional wisdom would have to conclude that these results indicate that we do not have free will. Let's take a particular experiment to see why...

"In one experiment, students were shown a list of words and then asked to recall words from it, after which they were told to type words that were randomly selected from the same list. Spookily, the students were better at recalling words that they would later type."[24] Therefore, if students could recall words better before the causative event even happened, then that seems to imply that they are not really in control of their choices and hence have no free will. The odds against chance for such experimental results is beyond 1 in a million.[25]

However, Digital Consciousness comes to the rescue and offers not one, not two, but three different free-will-friendly explanations for these results.

1. Evidence is rewritten after the fact. In other words, after the students are told the words to type, the RLL goes back and rewrites all records of the students' guesses, so as to create the precognitive anomaly. Those records consist of the students' and the experimenters' memories, as well as any written or recorded artifacts. Since RLL is in control of all of these items, the complete record of the past can be changed and no one would ever know.

2. RLL selects the randomly typed words to match the results, so as to generate the precognitive anomaly.

3. We live in an observer-created reality and the entire sequence of events is either planned out or influenced by intent, and then just played out by the experimenter and students. The precognitive feeling that the students may be experiencing could be owing to the fact that the trigger words given after the students' selection have already been picked by RLL. Via the "leaky data" concept (see "Paranormal—Explained" prior), this information is subliminally presented to the consciousness of each subject. Given the rest of the evidence for Digital Consciousness, this is the most likely explanation.

Mystery solved, Digital Consciousness style.

Metaphysical Anomalies—Explained

One in five American adults say that they have seen or been in the presence of a ghost. And 29 percent believe that they felt in touch with someone who had already died.[26] In a recent *National Geographic* poll of Americans, about 10 percent claim to have seen a UFO and a total of 36 percent believe that they exist.[27] In a CNN Poll, 50 percent of the respondents believed that extraterrestrials have abducted humans, with 7 percent claiming that it happened to them or someone they know.[28] Results are similar in many countries around the world. In Iceland, a country with a far higher literacy rate (99 percent) than the US (86 percent per the Census Bureau), 54 percent believe in elves.[29] It is estimated that 11,000 people have seen what they believe to be the Loch Ness Monster.[30]

Now, I am not claiming that any of these things exist in our normal reality. But doesn't the fact that millions of people in highly educated countries have seen anomalous entities bear consideration? Presidents of the US (Reagan and Carter), the Associate Minister of Defense in Canada (Paul Hellyer), astronauts, high-level military officers and many reputable scientists have all reported UFO sightings. Furthermore, in many cases, these anomalous objects often behave in a manner that defies law of aerodynamics and/or physics.

Lee Katchen, a NASA atmospheric physicist, made an announcement in 1968 concluding that, after his examination of 7000 UFO reports, UFOs have an extraterrestrial origin. He stated:

> UFO sightings are now so common, the military doesn't have time to worry about them… when a UFO appears, they simply ignore it… Unconventional targets are ignored because apparently we are only interested in Russian targets, possibly enemy targets. Something that hovers in the air, then shoots off at 5,000 miles per hour, doesn't interest us, because it can't be the enemy. UFOs are picked up by ground and air radar, and they have been photographed by gun camera all along. There are so many UFOs in the sky that the Air Force has had to employ special radar networks to screen them out.[31]

So what is going on?

A very plausible explanation is provided by the Digital

Consciousness model. Per Figure 8.18,the vast majority of objects in RLL adhere to known (RLL) laws of physics:

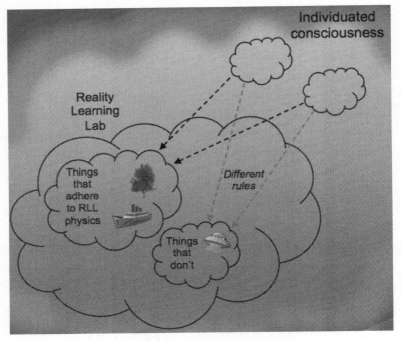

Figure 8.18

We all interact with these things and receive the same data about them so as to create the apparent objective consensus reality that we all know and love. However, it is perfectly feasible to have a category of objects that:

1. Do not necessarily behave according to "normal" laws of aerodynamics and/or physics.
2. Do not follow rules of consensus reality.

So, for example, imagine that the RLL UFO object is one such entity. First, it should be noted that nothing that has been reported really defies known laws of physics; rather, they merely defy what is considered possible with today's technology. Even if an object were to appear or disappear suddenly, this is well within our technological sights in terms of cloaking techniques. Harder to explain conventionally, however, is the anomaly that only certain people see these things. While there are plenty of consensus experiences where many people see the same anomalies (for example, well known UFO sightings such as the Rendlesham Forest incident or the Phoenix Lights), there are also cases where specific individuals see things that others don't. Even this finds a perfect explanation with our theory.

Imagine an MMORPG that is programmed to present consistent sensory data to all players. However, there may be anomalous objects in the game that don't follow the rules that everything else does. Moreover, the properties of those objects that are experienced (their color, speed or very existence) may be a function of the player's skill, strength and/or magic level.

In an analogous manner in ATTI, the experience that each IC has regarding an object in RLL may be a function of the specific IC itself. Again, it is impossible to know why ATTI evolved that way or why such patterns exist, except that they probably have some value in assisting certain conscious entities in their learning journey.

Former radio broadcasting executive and consciousness researcher Robert Monroe encountered all sorts of entities in his astral travels, including some who had never had the RLL "Earth" experience. It seems that these entities had a different path for their souls and it makes one wonder if ATTI is simply experimenting with a variety of methods for consciousness evolution. Monroe did not experience physical entities from a non-Earth physical environment. But, of course, thousands of "abductees" have (again, recall that many of his experiences were corroborated by hard evidence and consensus experiences of others).

The Mandela Effect—Explained

Recall that RLL is the physical universe that we know and love, but it is actually virtual. This virtual world is all a subjective experience of our true consciousness, which sits somewhere as part of ATTI. Hence, ATTI can modify our virtual world, as could another conscious entity within ATTI (who perhaps has an evolved level of access). I'm not sure which of these is messing with the historical artifacts, but either is entirely possible. It would be analogous to being a programmer of a multiplayer virtual reality fantasy game, and deciding to go back into the game and replace all of the pine trees with palm trees. The players would certainly notice, but they would think that there was a patch applied to the game for some reason and wouldn't really give it a second thought, because they realize the game is virtual. The only reason the Mandela effect freaks us out when we discover one, like Dolly's braces, is because we don't realize our reality is virtual.

As I write this, it feels like I am documenting something significant. However, I realize that tomorrow, this section of the book may be gone. Or perhaps the references that I listed to Dolly with braces will have disappeared and along with them, the original sources. And closed-minded science snobs like Bill Nye and Neil deGrasse Tyson will say it always was that way.

The flaw is in the assumption that "we" are all in the same reality. "We," as we have already discussed, are experiencing a purely subjective experience. It is the high degree of consensus between each of us "conscious entities" that fools us into thinking that our reality is objective and deterministic. Again, quantum physics experiments have proven beyond a reasonable doubt that it is not.

So what is going on?

Remember that "we" are each a segment of organized information in ATTI. RLL, what we experience every day while conscious (while meditating, or in deep sleep, we are connected elsewhere) is where all of the artifacts representing Jaws and Dolly exist. It is where various "simulation" timelines run. The information that represents our memories is in three places:

1. The "brain" part of the simulation. Think of this as our cache.
2. The temporary part of our soul's record, which we lose when we die. This is the stuff our "brain" has full access to, especially when our minds are quiet.
3. The permanent part of our soul's record; what we retain from life to life, what we are here to evolve and improve, what in turn contributes to the inexorable evolution of ATTI. Values and morality are here. Irrelevant details like whether or not Dolly had braces don't belong.

For some reason, ATTI decided that it made sense to remove Dolly's braces in all of the artifacts of our reality (DVDs, YouTube clips, etc.). But, for some reason, the consciousness data stores did not get rewritten when that happened and so we still have a long-term recollection of Dolly with braces.

Why? ATTI just messing with us? Random experiment? Glitch?

Maybe ATTI is giving us subtle hints that it exists, that "we" are permanent, so that we use the information to correct our path.

We can't know. ATTI is way beyond our comprehension.

Dark Matter—Explained

"Any sufficiently advanced technology is indistinguishable from magic."
- *Arthur C. Clarke*

In the late 18th century, the prestigious French Academy of Sciences believed that it was impossible for rocks to fall from the sky and, hence, reports of meteorites must be false.[32] According to author Jonathan C. Smith, "museums throughout Europe tossed out their meteorites as superstitious rubble."[33]

In 1967, astronomers Jocelyn Bell Burnell and Antony Hewish discovered pulses of radiation that were extremely consistent and continuous at 1.33 seconds between each pulse. Dubbed pulsars, nobody had any idea at the time what was causing them and there was of course plenty of speculation about an intelligent origin.

Today, we understand meteors and rapidly spinning neutron stars as the true causes for these anomalies. But, of course, we always have new anomalies to figure out.

There are always elements of the RLL that we don't know about. As we discover ways to experience those things, they become anomalous. Through experimentation, we discover the rules that govern their behavior and they become part of the status quo in physics, and are therefore no longer anomalous. A current case in point is dark matter.

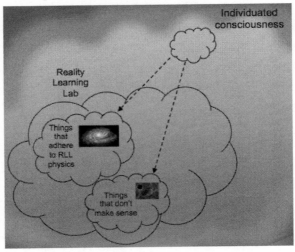

Figure 8.19

However, this also begs a "chicken-or-egg" question. Are the "things that don't make sense" part of RLL prior to our discovery of them? Or are they dynamically generated, in the same sense that there is no objective reality and that consensus subjective reality is created by consciousness (as in the case of the double-slit experiments)? For the purposes of this chapter, it is sufficient to simply explain the existence of these anomalies. But it is also interesting to dive a little deeper and see if there are any clues to this chicken-or-egg question...

Last Thursdayism and Symbiotic Evolution

The answer might come from our old friend, evolution, or the fundamental law of continuous improvement. Given that ATTI evolved according to such a law and that it evolved the RLL according to such a law, which would be more likely: that RLL would evolve first, and therefore be sophisticated enough for conscious entities to begin to play in? Or that RLL would evolve as IC evolved?

In reality, there might be a spectrum of evolutionary possibilities, with symbiotic evolution of IC at one end of the spectrum and Last Thursdayism at the other end. Last Thursdayism is the idea that the world started last Thursday, with all necessary artifacts (fossils, newspapers, our memories) neatly in place, to infer a complex and rich history of the universe and our existence. Of course, Last Thursdayism is just a metaphor — any day will do.

Last Thursdayism would look like this:

ATTI evolved RLL independently of ICs. Maybe the ICs spent a near eternity interacting with each other in ATTI space, frustrated with their inability to eat Dodger Dogs and have great sex. Perhaps even a variety of RLLs were attempted over the eons, not unlike the various Matrices made by The

Architect in *The Matrix* trilogy. Finally, we have RLL 42.0, our current instantiation, largely free of defects. A lottery system gives all of the eager-to-experience-reality ICs a chance at the new learning lab. Opening day—next Thursday. And so reality begins, say, on 14 March 1879 (Einstein's birthday, which was actually a Friday).

In this case, dark matter and other "things that don't make sense" are already there lurking in the dark programmatic recesses of RLL, fully formed, and waiting to be discovered, modeled, hypothesized and theorized. IC Karl Popper comes up with the scientific method, and left-brained ICs evolve their consciousness by following a path of gradual scientific and technological epistemological evolution. Relativity, quantum mechanics, dark matter, cold fusion, teleportation, utility fog.

Symbiotic Evolution would look like this:

Early RLL is a mess. It's buggy, grainy and the color blue doesn't exist.[34] ICs are tiny and barely able to regulate a gene. But part of their template is to have an experience in RLL, so they connect and live out largely meaningless lives as humble Archaea. The law of continuous improvement finds that larger multicellular templates in RLL provide even better opportunities for IC learning, so ICs evolve slightly in complexity and connect to such templates in RLL. And so on until, presto-change-o, modern humans, pyramids and Duck Dynasty.

Invoking Occam's razor, I'm going with Symbiotic Evolution.
Which means that we are probably developing the dark matter anomaly as we speak.
Isn't philosophy fun?

Multiple Dimensions, Parallel Universes, Parallel Realities — Explained

A topic like this deserves not just its own book, but also perhaps an entire library. Or a parallel array of libraries. Parallel universes and parallel realities are concepts that have been in existence since time immemorial. Buddhist philosophy refers to 31 planes of existence. Heaven and Hell are otherworldly realms in the Christian theology. Ancient Hindu texts, called the Puranas, refer to "countless universes, each enveloped in its shell."[35]

The problem is that terms like "multiple dimensions," "parallel universes," and "parallel realities" are tossed about in a wide variety of contexts and mean different things to different people. Cosmologists tend to think of multiple universes in a physical sense — bubbles of four-dimensional space-time, between which it is impossible to travel owing to constraints of physics. Quantum physicists might think of multiple universes in a quantum decision sense — the Many Worlds Interpretation (MWI) of quantum mechanics theorizes a multiverse consisting of an unaccountably large number of parallel realities, basically one for every instance imaginable, in a "reality" space called Hilbert Space. Mathematicians may think of multiple universes in a mathematical sense — multiple physical dimensions that allow for parallel realities across the unseen dimensions. And we digital philosophers might think of multiple universes more in a virtual sense, not unlike how different groups of avatars can occupy different virtual realities running on the same computational construct.

Let's take each of these ideas in turn and see what makes sense in terms of digital consciousness theory.

Cosmological multiverse

We used to be taught that the physical universe was everything there is. But, over the past few decades, new theories about the early universe has led to the idea that even physical reality may extend beyond the observable universe. Here's why...

Despite the possibility of having the most advanced equipment to peer into the deepest depths of space, there will always be a "practical observational horizon," beyond which the (apparent) laws of physics precludes us from seeing further owing to the speed of light. (I say "apparent," because history has shown that even laws of physics were meant to be broken.) At this theoretical observational distance limit, also known as the Hubble volume, the light from objects has been traveling toward us since the beginning of the Big Bang. Our Hubble volume is generally accepted to be about 42 billion light years in diameter. But that doesn't mean there's nothing beyond it. In fact, the most generally accepted cosmological theory—physicists Andre Linde and Alan Guth's inflationary Big Bang theory—allows for quite a bit of the material from the Big Bang to exist beyond our Hubble volume, because the inflationary period was superluminal. (What was I just saying about the speed of light?)

Linde takes this idea of matter beyond the Hubble volume to a different level. He notes that owing to the theoretical behavior of quantum fields, the inflationary expansion was chaotic in the sense of proceeding at different rates in different regions. As a result, it was likely that the universe expanded in a fractal manner, with bubbles of inflation sprouting other bubbles of inflation, each one being a new "Big Bang." This theory actually has far-reaching implications. If true, the universe, or multiverse, is essentially immortal, continuously evolving and growing, as shown in Figure 8.20.

"As the bubbles in the water well up and disappear again, so is the Universe created."
- *Guru Nanak, founder of the Sikh faith*

Figure 8.20

In fact, the question of a starting point to the multiverse is almost irrelevant, as it cannot be verified one way or the other owing to the fact that our Hubble volume is just within our local bubble.

Cosmologist Max Tegmark has described some real evidence for such a multiverse. Specifically, he notes that recent studies on the distribution of galaxies and the uniformity of background radiation demonstrate that matter in the universe is uniform out to at least 10^{24} meters (roughly 1/400th of the Hubble volume). As a result, it is highly likely that beyond our Hubble lies more of the same.

An interesting artifact of the cosmological multiverse theory is the realization that if enough of these separate bubbles of matter exist, eventually there must be another Hubble volume identical to ours. In fact, it can be estimated mathematically that, on the average, about $10^{10^{118}}$ meters away, there is an identical Hubble volume; for example, our "alter universe" (and, of course, Bizarro Universe would have to be about the same distance). If you want to find the nearest identical copy of you, statistically, it would about $10^{10^{28}}$ meters away.

But, other than the homogeneity data that implies a somewhat larger than Hubble volume universe, there is no other real evidence that there is a physical multiverse where different yet similar states of reality exist.

If other bubble universes exist, would it even make sense to speculate about them if it is impossible to visit them? Remember what we are talking about here. The "physical" multiverse, even if it includes space-time beyond the Hubble volume, is actually virtual. It is simply an extension of the RLL that we are now beginning to explore. And since our consciousness is elsewhere, it should actually be possible to visit the rest of the multiverse; although not in the "physical" sense. Anecdotal evidence supports the idea that meditation and in-between lives offer mechanisms to explore what lies beyond our normal concept of the "Big Bang" universe.

Figure 8.21

MWI multiverse

In Chapter 5's ridiculously concise primer on quantum mechanics, we briefly discussed the myriad interpretations of QM that have been put forth throughout the years and noted that one of the most popular is Hugh Everett's MWI. MWI asserts that every time a quantum mechanical decision is made (for example, does the particle go this way or that way, decay or not decay, interact with another particle or not?) a new universe is created, so that each decision is valid in one of the two. Figure 8.22 depicts MWI graphically. The exponential universe forking ad absurdum rapidly leads to an unimaginably large number of universes, consisting of every conceivable state, that make up a "universe space" called Hilbert space. Conventional (an oxymoron when applied to this theory) wisdom holds that it is

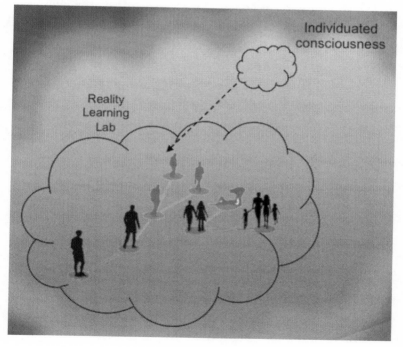

Figure 8.22

impossible to travel between any two elements of Hilbert space. But even this has been argued against by the likes of Rainer Plaga, who speculates that communication may actually be possible between the universes.[36]

I have declared that this theory is the most egregious violation of Occam's razor ever conjured. Esteemed physicist Paul Davies appears to agree and has described MWI as "cheap on assumptions, but expensive on universes."[37]

While there is no direct evidence for the MWI theory, many physicists have taken a crack at hypothesizing tenuous supporting evidence, but it pretty much all falls in the category of theoretical experiments that are far out of the reach of today's technology or the results of a quantum suicide experiment. In a quantum suicide experiment, the subject's life depends (like that of Schrödinger's cat) on the measurement of a property of a particle that can change over time. If the subject perceives that he or she never dies, then MWI is likely true. The problem with the experiment is that all external observers will see the same result and even if we find a willing Kevorkian-esque participant as the subject, we would never be able to communicate with them should they branch, MWI-style, into another universe, so no one who survives the experiment can record its results for the rest of us.

Multi-dimensional multiverse

Flatland: A Romance of Many Dimensions was a book written in 1880 by Edwin A. Abbott. In it, he describes a two-dimensional world, called Flatland, inhabited by two-dimensional creatures, all blissfully unaware that their world is really part of a three-dimensional one.

And just as the Flatlanders can be unaware of a larger-dimensional universe around them, so might we be unaware of a larger-dimensional universe. Neither physics nor mathematics precludes such possibilities. And some argue that dark matter may be evidence for the existence of higher dimensions.

In *The Universe—Solved!*, I wrote about brane cosmology, the theory that we live on the "wall" of a higher-dimensional space, or brane, made possible by the ideas of string theory. The higher dimensions that we live in, or possibly other nearby parallel branes, can theoretically exert an influence on our 3D space in the form of gravity, which may be the explanation for dark matter.[38] I noted that the LHC (Large Hadron Collider) might be able to validate or invalidate such ideas within a few years of the publication of that book. Well, the results are in and, as always, they are inconclusive, in that they set limits on the possible sizes of microscopic black holes in extra-dimensional universes and therefore limit the possibilities of higher dimensions. This evidence tends to argue against such models, but certainly not conclusively.

The dimensions that we are talking about here are spatial dimensions. Figure 8.23 shows how higher-spatial dimensions

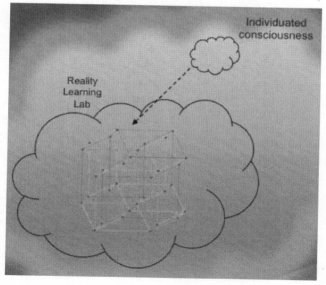

Figure 8.23

may be possible. Our three-dimensional space defines a position by three coordinates, each of which establish a measurement in a direction that is orthogonal to the other two. These coordinates can be thought of as being discrete points in space. But imagine that there is a completely identical spatial construct right next to the 3D space that we know and love—just that you can't see it or get to it because we can't move in that dimension. Some people wonder if dark matter is just the effect of mass being very close to our 3D world in that higher dimension. If so, it might be that gravity can bleed over across this higher dimension while other forces, such as the electromagnetic force responsible for the transmission of light, cannot (and hence the invisibility of dark matter).

However, it is important to consider what is meant by "dimension." The discussions around higher dimensions usually refer to spatial dimensions. But the multiverse theory gets much more interesting if you relax the definition of *dimension*.

Multi-virtual multiverse

A dimension in the broader sense is just a defining variable or a means of categorization. In terms of the attribute of color, for instance, there are generally considered to be three dimensions: hue, saturation and brightness. We use the idea of three-dimensional space to define an absolute position within it; for example, to identify the position of an object. We add the fourth dimension of time to identify the time at which that object exists at that position. But to fully describe the object itself, we would need many more dimensions, such as the three dimensions of color.

In MMORPGs, a given server (computer that serves up the game to the players) can only handle so many players at once. So, traditionally, there are multiple servers, each of which provide a virtual reality experience for a subset of the entire population of players of that game. This means that when you log in, you are in an instance of the game along with many other players, but there are other instances of the same game going on at the same time with other sets of players. This may be changing with more scalable architectures, but the concept is a perfect example of a multi-virtual reality.

Figure 8.24 below shows how it may be applied in the Digital Consciousness context. ATTI is arbitrarily large and complex, and may certainly have evolved to run multiple learning labs at the same time. Perhaps this is why the mystics and experiencers all speak of everything happening simultaneously in the afterlife. Perhaps there is not a linear progression of lives, but rather, for a given IC, some may proceed in parallel. In some sense, given that the higher objective is to learn and evolve, it may be faster to do it in parallel, assuming that the IC can actually "time slice" in this manner. On the other hand, it would seem that one could mostly benefit from and build on previous learnings if they are sequential. Still, the model shows how flexible ATTI and RLL may be.

There is actually considerable anecdotal evidence supporting temporary anomalous time and space shifts. Like the time when two well-respected professors from St. Hugh's College in Oxford, UK paid a visit to the Palace of Versailles and came across Marie Antoinette sketching in a garden, along with a number of others in 1780s period attire. Marie and her entourage disappeared when a tour guide approached in "real time." Or the policeman who was shopping with his wife in London in 1996 and stepped into the 1950s—cars, clothes and even a clothing store named Cripps, which hadn't existed since the 50s—before reemerging in 1996.[39] Or the time that RAF Air Marshal Sir Robert Victor Goddard inspected and confirmed an abandoned airfield in Drem, Scotland in 1935. After taking off and returning owing to bad weather, he found a bustling airfield with mechanics in blue overalls (which he thought unusual because mechanics wore brown overalls) and yellow training planes that he couldn't identify because the RAF had no yellow planes at the time. Four years later, his vision came to pass as Drem airfield was in operation, complete with mechanics in blue overalls and yellow training planes called Miles Magisters.[40]

These and other temporary space and time shifts can be fully explained with digital consciousness. The spatial shift is simply the consciousness temporarily interrupting its experiential flow, and jumping to another locale in the RLL and back. A time shift to the past is also straightforward. Considering that all experiences can be recorded in ATTI, it would not be difficult for the consciousness to insert itself into a part of the recording in the past, especially if it is merely observing and not interacting with the reality. Even if interacting, it could be done in a parallel virtual instance of the reality without violating the grandfather paradox (the time-travel paradox whereby you travel to the past and kill your grandfather, thereby making your own existence impossible).

You are experiencing your current reality in Reality Instance A and experience a time shift to the past, but your consciousness is actually connecting to Reality Instance B. You can interact all you want in B, because it won't impact the apparent later time when you return to A. The same holds for a time slip into the future. The future may already have happened in your instance (A) and in your current day reality, you are merely running through the experience, so that when you slip into the future you really are seeing what happens. Alternatively, you may jump to another instance of reality (B), one in which events play out in a manner very similar to the way they will ultimately play out in your current instance (A).

In these ways, time travel anomalies are entirely explainable. Time travel itself is possible; although the reality instance may be different.

These anomalies, while anecdotal, are still bona fide evidence for the existence of a multi-virtual multiverse.

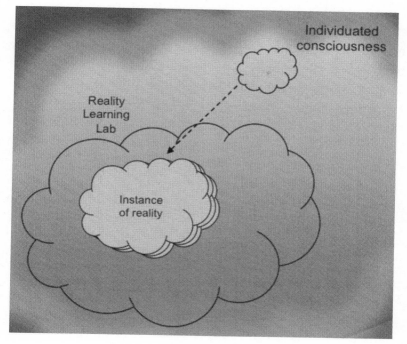

Figure 8.24

Other Anomalies — Explained

Slime mold

In Chapter 6, we explored the mystery of how slime molds can perform complex learnings without having a nervous system.

The informational substrate in which consciousness resides is either "the true physical reality" or a "truer reality" than the virtual one in which we think we reside. It is that substrate that may contain the complexity for memory and learning on the part of the consciousness of the organism.

For my Masters project in college, I had to develop a system that would take rich complex information from weather balloon sensors and crunch the data to match the low bit rate telemetry limitations of the transmission system. In an analogous manner, perhaps, the consciousness that got stuck with the poor slime mold template has very little to work with in terms of interacting with its virtual world. But all that the mold really needs is a small subset of the three learning process elements described in Chapter 6: the ability to sense and deliver information to its conscious host, and the ability to respond to instructions from that host and interact with its environment. The consciousness does the rest.

In a similar way, a perfectly respectable IC may be stuck with a cognitively challenged human template running as a presidential candidate.

Xenoglossy (foreign accent syndrome)

Past-life regression hypnotists occasionally come across an emerging persona who speaks a language completely unknown to the patient, ostensibly from a past life. Famous US psychic Edgar Cayce himself spoke fluent Italian on one occasion while in a trance; although he had never known the language. Ian Stevenson has researched and documented this phenomenon extensively. To cite one example, a hypnotist named Carroll Jay hypnotized his wife once in 1970 to help her with back pain and a German persona named Gretchen emerged. Stevenson investigated the case and brought three native German speakers to witness her, two of whom were also doctors. Over a number of recorded sessions he documented hundreds of German words that she introduced into the conversation (not repeated from questions asked by the German witnesses). The persona Gretchen also identified the family, town and life events from her past. And yet, she had never learned any German, at all.[41]

These cases seem most mysterious until one considers that the consciousness has access to a historical record of past lives and hence, knowledge of native languages from past lives via the permanent store of data in ATTI. Meditation, hypnosis, ritualistic dance or drumming, trauma and sometimes pharmaceuticals can bring on the access to this data. And sometimes it occurs for no apparent reason.

Chapter 9

The Final Proof

In Chapter 4, we used Venn diagrams to illustrate the relationship between questions or anomalies and the theories that address them. The diagrams can either be sets of theories that can explain a particular anomaly or sets of anomalies that are well addressed by a particular theory. In either case, the diagram provides us with a very scientific process of abductive analysis applied to theories of reality.

Let's see how Digital Consciousness Theory stacks up against some of the others out there. Here is our competition:

1. Abrahamic religions—God created everything.

2. String theory—Everything is ultimately composed of tiny vibrating strings. String physicists claim that string theory unifies quantum mechanics and relativity, despite its uncomfortable requirement of 10 or 26 dimensions. This would be a subset of the general category of deterministic materialism.

3. Deterministic materialism—This is the general idea that everything is physical, objective and deterministic. I include this to make it ridiculously clear how little this ancient paradigm actually explains.

4. MWI and Anthropic principle—Some physicists and mathematicians hang their TOE hat on the Everett interpretation of quantum mechanics which, although it has explanatory power as it related to quantum mechanics, it neither passes Occam's razor nor an informal "sniff test." I couple it with the Anthropic Principle, because only that pairing can explain the finely tuned universe.

5. Simulation theory—Here I am referring to the more traditional idea that we live in a simulation, à la *The Matrix*,

which was written by another entity.

6. Eastern philosophies—In particular Hindu and Buddhist philosophies.

And we shall add Digital Consciousness as TOE #7 (of course!).

And let's analyze the effectiveness of these theories on the anomalies that we have discussed in the book. In some cases, they can probably be grouped together—for example, quantum Zeno and quantum entanglement would probably share the same TOEs. Paranormal and metaphysical anomalies would probably also have the same TOE explanatory scorecard. So, I will select the following as unique and fairly orthogonal anomalies. Certainly, the one that can explain them all deserves serious recognition as the *ultimate* TOE, right? Let's see what we can explain these anomalies with...

1. Quantum entanglement and quantum Zeno
2. The finely tuned universe
3. NDEs
4. The Mandela effect
5. Paranormal and metaphysical
6. The nature of matter
7. Unification of relativity and quantum mechanics—I include this to give string theory a *raison d'être*.
8. The observer effect
9. Meditation, the placebo effect and the power of positivity—all of these things are artifacts of the same effect.
10. Nature vs. nurture
11. Delayed-choice quantum eraser
12. Precognition
13. Spiritual experiences
14. Dark matter

First, let's look at TOE space and see which TOEs are satisfied by each of these anomalies. Because we are looking at so many different anomalies, it is much easier to visualize these relationships by breaking them into two different Venn diagrams, as show below. Here is the first set of anomalies:

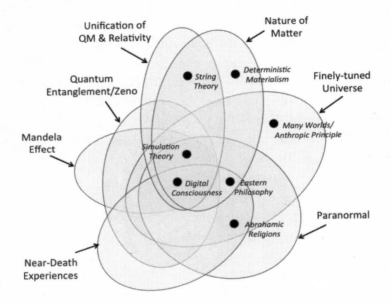

Figure 9.1

For example, from the diagram, you can see that there are five TOEs that have explanations for the nature of matter: string theory (matter is vibrating strings), deterministic materialism (matter might be vibrating strings but could also be based on standard particles), simulation theory (matter is information), Digital Consciousness (ditto) and Eastern philosophy (matter is ephemeral vibration). The MWI theory and Abrahamic religions really have nothing to say about the nature of matter. One could argue that Abrahamic religions do have an explanation—that God created matter—but it doesn't explain what matter is fundamentally. Each time science progresses, the same argument progresses in lockstep; for example, God created atoms, then He created subatomic particles, then He created strings, etc. So Abrahamic religions don't add any real critical value to an explanation for the nature of matter.

Some things to note from this diagram:

1. Deterministic materialism doesn't explain much, only one mystery of physics.
2. The MWI theory doesn't explain much, only the finely tuned universe.
3. Only the simulation and Digital Consciousness theories can explain quantum entanglement, quantum Zeno and the Mandela effect.
4. Only Digital Consciousness sits at the intersection of all of the anomaly sets, therefore having an explanation for all seven of these anomalies.

Here is the second set of anomalies:

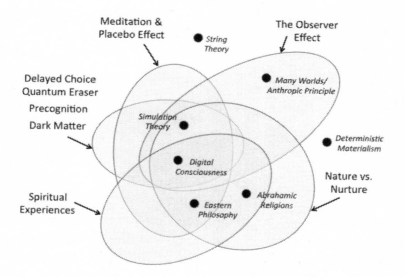

Figure 9.2

Looking at the second set of anomaly sets, we can note a couple more interesting things:

1. String theory doesn't explain any of these seven anomalies.
2. Deterministic materialism doesn't explain any of them, either.
3. Again, only Digital Consciousness sits at the intersection of all of the anomaly sets, therefore having an explanation for all seven of these anomalies, as well.

Let's now look at anomaly space and see which anomalies have explanations from each TOE:

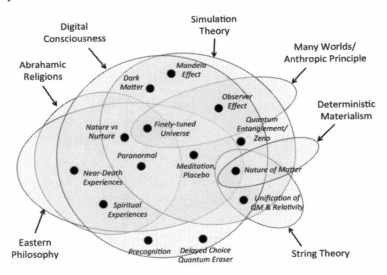

Figure 9.3

From this diagram, it is very easy to see how many different anomalies each TOE can explain. Let's rank them according to their explanatory power (by number and percent of anomalies explained):

1. Digital Consciousness theory—14 anomalies (100 percent)
2. Simulation theory—10 anomalies (71 percent). Simulation theory of *The Matrix* variant doesn't make a lot of sense in terms of explaining spiritual experiences like NDE or general mystical spiritual experience. Also, it is difficult to find a rationale in deterministic simulation theory for reverse causality anomalies like precognition and the delayed-choice quantum eraser.
3. Eastern philosophy—7 anomalies (50 percent). Hindu and Buddhist philosophies actually have excellent explanatory power for most things related to the nature of our universe and our place in it. But it has little to say, of course, on the quantum and general physics mysteries of our time.
4. Abrahamic religions—5 anomalies (36 percent)
5. MWI/anthropic principle—2 anomalies (14 percent)
6. String theory—2 anomalies (14 percent)
7. Deterministic materialism—1 anomaly (7 percent)

By this methodical analysis, it should be clear that deterministic materialism is as dead as a doornail. As dead as disco. As dead as Michael Vick's career. MWI/anthropic isn't far behind and has one foot already in the grave. This is because it was dreamed up to explain a couple particular anomalies—the finely tuned universe and the observer effect, without invoking a role for consciousness. String theory isn't really dead, but its explanatory power is limited to the nature of matter and the unification of two other theories of physics.

In summary, it should be clear that only digital consciousness theory can offer explanations for many if not all of the world's mysteries, both scientific and metaphysical. Therefore, by the logical method of abduction, I submit that Digital Consciousness Theory *is* the true theory of everything.

Chapter 10

Making Sense of Everything Else

"The universe begins to look more like a great thought than a great machine. Mind no longer appears to be an accidental intruder into the realm of matter. We are beginning to suspect that we ought rather to hail it as the creator and governor of this realm."
- *James Jeans*

So this very powerful idea, Digital Consciousness, seems to explain or provide a consistent framework for the explanation of virtually every mystery known to man. What does that say about everything else? Are ideas like the holographic paradigm, the power of positive thinking or string theory, without merit? Actually, not at all. Many of the ideas of great physicists like David Bohm, Brian Greene and Michio Kaku are very sound. Popular New Age philosophies from writers like Rhonda Byrne (*The Secret*) and Malcolm Gladwell will find Digital Consciousness to be a very suitable foundation. Ancient wisdoms and prophets weren't wrong, either, as many of the ideas from Hindu, Sikh, Buddhist, Christian and Islamic holy books fit like a glove. In fact, if anything, Digital Consciousness is like the great unifier, making sense not only of the discoveries and theories of modern science, and the wisdom of the ancients, but also clarifying the reasons for their limitations.

Spiritual Experiences, Cellular Automata and Everything In-between

Consider the following (by no means comprehensive) hierarchy:

Spiritual Experience	Newtonian Physics	
	Grand Unified Theory	
Placebo Effect, Power of Positive Thinking, Visualization	Relativity	Quantum Mechanics
Paranormal	String Theory	
Holographic Theory		
Cellular Automata		
Digital Consciousness Theory		

Figure 10.1

The table is divided into three columns only so that relativity and quantum mechanics can have their proper place. Both of those theories are incompatible, and yet have an excellent track record for experimental support and prediction. Therefore, they must be valid as models of our reality at some level or with some constraints. In the relativity case, it is a more fundamental description of the mechanical physics than Newtonian, because it incorporates effects only seen at high speeds. Newtonian physics is a great model at the macroscopic and slow scales, two significant constraints. In other words, it requires a deeper model of reality to cover effects at high speeds and at very tiny scales. Quantum mechanics itself is just a model and obviously can't explain everything, as Chapter 9 made abundantly clear. To explain quantum anomalies requires a deeper model, such as string theory.

Yet, string theory is itself just a model. As we have seen, there isn't much that it explains. While it unifies QM and relativity in some cases, it is silent on entanglement and other quantum anomalies. It also has the obvious logical flaw that since consciousness seems to be able to position matter in different places depending on intent, matter can't really be fundamentally material. However, even string theory predicts a material model for matter. This adds no value to the model. If each particle in string theory only differs by the frequency of the vibration of the string, what do you need the material for? The frequency is just data. And data can easily be manipulated by data. And the mystery of the observer effect disappears.

So, underneath string theory must be something else, which can model the unexplained effects of quantum mechanics. Before we get to it, let's have a look at the other family of reality experiences — the spiritual, mystical and metaphysical.

It is a complete and total mistake to discount the validity of spiritual experiences. All that we have are experiences. And for the history of modern humanity, there is evidence of spiritual experience in all cultures across the globe. It didn't end in the so-called age of reason either, as people throughout the world commonly have spiritual experiences today. But those experiences also represent a model. We look for common aspects to the reports of such experience that give us an indication of the nature of a truer reality. What we come up with is life after death, being all connected to a greater whole, the idea of God or ATTI, a separate consciousness, a purpose for our lives and reincarnation. This model is a little fuzzier than relativity, because it can't be expressed in nice neat equations and because it includes a strong element of probability. But, as it turns out, the same can be said for quantum mechanics.

Some explanatory layer down from spiritual experiences, we try to quantify scientifically, concepts such as the placebo effect, the power of intent, the effects of visualization and neuroplasticity. These ideas are also very probabilistic, but are all very valid models that describe reality. The power of belief is real. It had been proven scientifically and supported by millions of anecdotal examples. But what explains it? We have to look deeper.

Subtle effects of the paranormal could be considered more fundamental than the power of belief. Or it could be models that represent different manifestations of effects and forces that we simply don't understand. However, research has shown that the effects are real yet probabilistic. What accounts for it? Again, we have to look deeper.

Holographic theory has been touted as a unifying concept for both physical anomalies as well as metaphysical ones. The effects of quantum mechanics that aren't part of the string theory model, like entanglement and the observer effect, find a logically consistent model in the holographic paradigm, which says that there is an underlying connectivity of all things, an *implicate order*. This is a good model and perfectly valid, but it can't be the final model, because the implicate order is left undefined. On the metaphysical side, it explains the power of belief and paranormal experiences, again owing to the interconnectedness of all things, including our brains, our cells and inanimate objects. So, yes, the holographic paradigm is an excellent model, albeit not rigorously formalized. Still, we must look deeper.

Some, like physicist Stephan Wolfram, have attempted to explain reality via the idea of cellular automata, which are essentially little FSMs comprised only of data elements. This could certainly be more fundamental than string theory, because it provides a basis for vibrating strings. If one applies multidimensionality to cellular automata, you could conceivably establish a substrate for the holographic paradigm — interconnectivity via a dimension that is invisible to us, while the mechanics of reality play out in our three experiential dimensions. No one has developed the theory that far, but that doesn't make it an invalid model of reality.

Even if someone were able to rigorously define multidimensional cellular automata that explains entanglement, could it explain NDEs or reincarnation? Nope. Only the fundamental level of reality can do that.

So it isn't until you get to the bottom layer that you can say you've hit reality. Of course, "Digital Consciousness" may also turn out to be a model, but it feels like it is distinctly different from the other layers. Meaning that, by itself, it is no longer an approximation of reality, but rather a complete and comprehensive yet elegantly simple framework that can be used to describe every single aspect of reality.

Transhumanism and Immortality — 21st-century Snake Oil?

Transhumanism is the idea that as science and technology progress, bionic enhancements will progressively become integrated with the human body until we effectively become a new species. Some refer to this projection of humanity as Humanity+ or simply H+. This movement is chock full of cool ideas, many of which make complete sense, even though they are perhaps obvious and inevitable. The application of science and technology to the betterment of the human body ranges from current practices like prosthetics and LASIK to genetic modification and curing diseases through nanotech. It is happening and there's nothing anyone can do to stop it, so enjoy the ride as you uplift your biology.

However, part of the transhumanist dogma is the idea that we can "live long enough to live forever." Live long enough to be able to take advantage of future technologies like genetic manipulation and uploading consciousness to silicon, thereby ending the aging process so "*You, too, can be immortal!*"

The problem with this mentality is that we are already immortal. And there is a reason why our corporeal bodies die. Simply put, we live our lives in this reality in order to evolve our consciousness, one life instance at a time. If we didn't die, our consciousness evolution would come to a grinding halt as we spend the rest of eternity playing solitaire and standing in line at the buffet. ATTI appears to evolve through our collective ICs. Therefore, deciding to be physically immortal could be the end of the evolution of the universe itself. Underlying this unfortunate and misguided direction of transhumanism is the belief (and, it is *only* that — a belief) that it is lights out when we die. Following that train of logic, if this were true, consciousness only emerges from brain function, we have zero free will, the entire universe is a deterministic machine and even investigative science doesn't make sense any more. So why even bother with transhumanism if everything is predetermined? It is logically inconsistent.

We take for granted that there is value to prolonging someone's life through technology and pharmaceuticals. But, could that actually be working against the fundamental purpose of ATTI? Consider...

Why We Die

The purpose of each instance of our soul's incarnation in this digital RLL is to evolve through learning and experiences. But, when we are first born, our digital self is incapable of absorbing any lessons and we have to first establish a basic virtual intelligence, which may take a few years. So, if you plot the consciousness quality for a single lifetime against time, it might look something like Figure 10.2. In fact, on average, our growth also probably tails off a bit, as shown, at the end of our lives, because we have absorbed all of the lessons that we can and all that is left to focus on is keeping the children off the lawn!

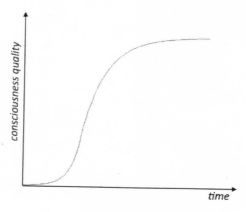

Figure 10.2

Now look at Figure 10.3. It depicts a case where a life is artificially extended beyond its natural duration. The end-of-life tail-off is much longer, as the subject spends more of their time yelling at the children on the lawn and not evolving.

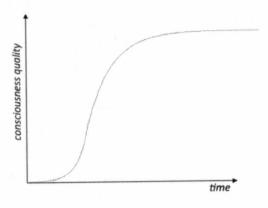

Figure 10.3

Since our consciousness evolution or "soul learning" will pick up in the next life where it left off in the previous one, the progressive evolution of consciousness can be depicted by stacking these curves, as shown in Figure 10.4:

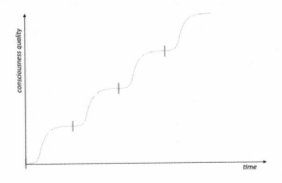

Figure 10.4

Note what happens when we plot the stack of natural life

consciousness evolution against the stack of artificially extended life consciousness evolution (see Figure 10.5):

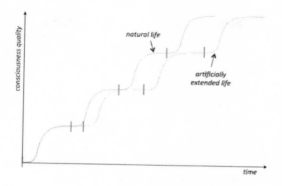

Figure 10.5

The natural life simply progresses faster than the artificially extended one. Another way of looking at this is to say that the natural lives will achieve a particular level of consciousness quality in a shorter period of time than the artificially extended ones. In fact, this is probably why we are programmed to die when we do. It ensures that our soul incarnates at a frequency that is optimal for the evolution of our consciousness. ATTI is efficient and has evolved RLL itself in order to be efficient.

This is not to say that it isn't worth watching out for your health and taking full advantage of "quality of life" extension opportunities (eating well, exercising, meditating, eating an apple a day, etc.) that are available to you. To extend a high-quality lifestyle still allows you to extend your opportunities to learn, love and grow in this lifetime. But I am not sure that the same can be said for uploading your consciousness into silicon.

Why We Live

It sounds like a conspiracy theory, but it rings true. We are controlled. The 40-hour workweek was designed to keep us busy enough to have neither the time nor the energy to question authority, but with just enough free time to be able to spend the money that we earn in order to keep the rich getting richer. The educational system, with its rows of desks, and emphasis on attendance and memorization, was designed to create obedient factory workers. In the US, our average standard of living has been in decline since 1973, while most of our tax dollars go to service war profiteers by creating enemies, and destroying and rebuilding their countries. Whistleblowers like economist and author John Perkins have explained with great clarity the way the system works—the corporatocracy, the financial pyramid scheme, the new world order.

As I write this, I have been watching a brilliant piece of historical democracy, the US Constitution, become progressively eroded over the past 15 years. Citizens bemoan their lack of choices in public elections, but continue to vote in the same lackeys with the same agenda, while preferring to be entertained by reality television and cat videos on Facebook (admittedly, much of that is incredibly entertaining!) rather than amass a meaningful movement against it all.

Why does all of this happen? Not because ATTI is evil. ATTI is agnostic to good or evil. Good and evil are simply human concepts, which are context specific. And humans are driven by free will and the information that is stored in their sticky soul data structures, plus the cache of the RLL. In the bigger picture, we progress over the long timescale. As just one instance of supporting evidence for this claim, consider how the murder rate in the EU has been steadily declining over the past 500 years and is less than 5 percent of what it was 500 years ago.[1] It might not feel like humanity is progressing when we see the horrors of modern war unfold. And, in some cases, we may be taking a step forward and then two steps back, instead of the other way around. But, in the end, we slowly progress. Why? Because each IC is on average progressing through learnings attained with each incarnation.

This is why we live. To learn, to evolve, to serve and to love. So that ATTI can evolve.

An appreciation for the true nature of reality, as has been proposed in this book, can help us to see this bigger picture. At a societal level, it can help us to recognize the futility of war and divisive behavior. It can help us to spot the politicians who are only in the game for their ego and not for any common good (I'm looking at you, Orange Lord). It can help us to direct public funds to the right causes. It can help us to focus on the quality of life instead of simply life extension.

At an individual level, it can reduce our fears. We don't have to fear death. There is no such concept as scarcity of resources driving a hedonistic game of survival. We can relax and follow our true callings, and respect our fellow humans and animals. We will also recognize that since our reality is "soft" and in a constant state of consciousness-driven creation, we can use that fact to drive improvements in our physical and mental health, relationships, career aspirations and life in general.

So put the stinkin' book down, and go out and live your life to the fullest.

Glossary

AI – Artificial Intelligence
ATTI – All That There Is
DCT – Digital Consciousness Theory
FSM – Finite State Machine
IC – Individuated Consciousness
MMORPG – Massively Multiplayer Online Role Playing Game
QM – Quantum Mechanics
RLL – Reality Learning Lab
VR – Virtual Reality

Notes and References

Chapter 2

1. Meyer, Stephen C. *Signature in the Cell: DNA and the Evidence for Intelligent Design* (New York: HarperCollins, 2009) p. 433.
2. Platt, Charles. "What if Cold Fusion is Real?" *Wired*, vol. 81, Issue 6.11. November 1998. Available at: <https://www.wired.com/1998/11/coldfusion/> [accessed 25 June 2017].
3. *New Energy Times*, Issue 16. 10 May 2006. Available at: <http://newenergytimes.com/v2/news/2006/NET16.shtml> [accessed 25 June 2017].
4. Plotkin, Marc J. "Cold Fusion Heating Up — Pending Review by U.S. Department of Energy," Pure Energy Systems News Service, 27 March 2004. Available at: <https://pesn.com/archive/exclusive/2004/ColdFusion_DOE/index.html> [accessed 25 June 2017].
5. CERN Colloquium on LENR, Geneva, 22 March 2012.
6. Taylor, Leonard. "A Gallery of Electromagnetic Personalities 2." University of Maryland, School of Engineering. 13 March 2016. Available at: <http://www.ece.umd.edu/~taylor/frame2.htm> [accessed 25 June 2017].

Chapter 3

1. Campbell, Thomas. *My Big TOE — The Complete Trilogy* (Huntsville, AL: Lightning Strike Books, 2007).
2. Available at: https://en.wikipedia.org/wiki/List_of_relativistic_equations [accessed: 25 June 2017].
3. Available at: <https://en.wikipedia.org/wiki/Theory_of_relativity> [accessed: 25 June 2017].
4. Campbell, p. 169.
5. Long, Jeffrey. "How Many NDEs Occur in the United States Every Day?" Near Death Experience Research Foundation, n.d. Web. 13 March 2016. Available at: <http://www.nderf.

org/NDERF/Research/number_nde_usa.htm> [accessed 25 June 2017].

6. Shermer, Michael. "What Happens to Consciousness When We Die." *Scientific American.* N.p., 1 July 2012. Web. 13 March 2016. Available at: <http://www.scientificamerican. com/article/what-happens-to-consciousness-when-we-die/> [accessed 25 June 2017].

7. Andrade, Edward Neville da Costa. *Rutherford and the Nature of the Atom* (New York: Doubleday & Company, 1964) p. 111.

Chapter 4

1. Bruce, Alexandra. *Beyond the Bleep: the definitive unauthorized guide to What The Bleep Do We Know!?* (New York: Disinformation Books, 2005) p. 33. (Note: the poll was published in the French periodical *Sciences et Avenir* in January 1998.)

2. Feyerabend, Paul. *Against Method* (London: Verso, 2010).

3. Available at: <https://plato.stanford.edu/entries/abduction/> [accessed 25 June 2017].

Chapter 5

1. Alamu, F.O. et al. "A Comparative Study on IFA Divination and Computer Science." *International Journal of Innovative Technology and Research.* Vol. 1, issue 6, October–November 2013, pp. 524–528.

2. Read, P. and Meyer, M. *Restoration of Motion Picture Film.* (Oxford: Butterworth-Heinemann, 2000) pp. 24–26.

3. Higgins, Chris. "Camera shoots at 4.4 Trillion frames per second." *Wired Magazine,* 12 August 2014. Available at: <http://www.wired.co.uk/article/worlds-fastest-camera> [accessed 25 June 2017].

4. Available at: <http://phys.org/news/2010-05-attoseconds-world-shortest.html> [accessed 25 June 2017].

5. Ali, A., Faizal, M. and Majumder, B. "Absence of an Effective

Horizon for Black Holes in Gravity's Rainbow." 8 June 2014. Available at: <http://arxiv.org/abs/1406.1980> [accessed 25 June 2017].

6. Tegmark, Max. "Parallel Universes." *Scientific American*, vol. 288, #5. May 2003: pp. 40–51.

7. Price, Michael Clive. "The Many-Worlds FAQ." February 1995. Available at: <http://www.anthropic-principle.com/preprints/manyworlds.html> [accessed 25 June 2017].

8. Available at:<https://en.wikipedia.org/wiki/Interpretations_of_quantum_mechanics> [accessed 25 June 2017].

9. Matthews, Jermey N. A. "Questions and answers with Amit Hagar." *Physics Today*. 14 January 2015. Web. 13 March 2016. Available at: <http://scitation.aip.org/content/aip/magazine/physicstoday/news/10.1063/PT.5.3019> [accessed 25 June 2017].

10. Brandenberger, Robert. "Introduction to early universe cosmology." *Proceedings of Science*. Available at: <https://arxiv.org/abs/1103.2271> [accessed 25 June 2017].

11. Dawid, Richard. *String Theory and the Scientific Method* (Cambridge: Cambridge University Press, 2013) p. 134.

12. Chang, L., Lewis, Z., Minic, D. and Takeuchi, T. "On the Minimal Length Uncertainty Relation and the Foundations of String Theory." *Advances in High Energy Physics*, 2011, 493514.

13. Hossenfelder, Sabine. "Minimal Length Scale Scenarios for Quantum Gravity." *Living Rev. Relativity*, 16, 2013.

14. Bostrom, Nick. "Are You Living In a Computer Simulation?" *Philosophical Quarterly*, 2003, Vol. 53, No. 211, pp. 243–255.

15. Bostrom, N., ref. 14.

16. Rincon, Paul. "Evidence of earliest human burial." BBC News, 26 March 2003. Available at: <http://news.bbc.co.uk/1/hi/sci/tech/2885663.stm> [accessed 25 June 2017].

17. Lee, Chris. "The dark side of light: negative frequency photons." *Ars Technica*, 24 August 2012. Available at: <http://

arstechnica.com/science/2012/08/the-dark-side-of-light-negative-frequency-photons> [accessed 25 June 2017].

18. Available at: <https://en.wikipedia.org/wiki/Electromagnetic_wave_equation> [accessed 25 June 2017].

19. Available at: http://worldafropedia.com/wiki/index.php?title=Adinkra_Symbols

20. Gates, James. "Symbols of Power." *Physics World*, June 2010.

21. Hogan, Craig J. and Mark G. Jackson. "Holographic Geometry and Noise in Matrix Theory." Physical Review D: Particles and fields, 79(12), December 2008.

22. Chown, Marcus. "Our world may be a giant hologram." *New Scientist*, 14 January 2009. Available at: <https://www.newscientist.com/article/mg20126911-300-our-world-may-be-a-giant-hologram/> [accessed 25 June 2017].

23. "That's odd: Weirdly energetic arrivals from outer space." *New Scientist*, 27 April 2016. Available at: <https://www.newscientist.com/article/2086129-thats-odd-weirdly-energetic-arrivals-from-outer-space/> [accessed 25 June 2017].

24. Whitworth, Brian. "The emergence of the physical world from information processing." Available at: https://arxiv.org/ftp/arxiv/papers/1011/1011.3436.pdf [accessed 25 June 2017].

25. Whitworth, Brian. "Quantum Realism." Available at: <http://www.brianwhitworth.com/BW-VRT1.pdf> [accessed 25 June 2017].

26. Whitworth, Brian. "Exploring the virtual reality conjecture." Available at: <http://www.brianwhitworth.com/FQXiWhitworth.pdf> [accessed 25 June 2017].

27. Tong, David. "Is Quantum Reality Analog After All?" *Scientific American*, December 2012, pp. 307, 46–49. Available at: <http://www.scientificamerican.com/article/is-quantum-reality-analog-after-all> [accessed 25 June 2017].

Chapter 6

1. Weihrauch, T.R. and T.C. Gauler. "Placebo-efficacy and adverse effects in controlled clinical trials." *Arzneimittelforschung*, 1999 May; 49(5): pp. 385–93. Available at: <http://www.ncbi.nlm.nih.gov/pubmed/10367099> [accessed 25 June 2017].

2. Kahn, Barbara E. and Alice M. Isen. "The Influence of Positive Affect on Variety Seeking Among Safe, Enjoyable Products." *Journal of Consumer Research*, September 1993, vol. 20, no. 2. Available at: <http://www.jstor.org/discover/10.2307/2489273> [accessed 25 June 2017].

3. Weizmann Institute of Science. "Quantum Theory Demonstrated: Observation Affects Reality." *ScienceDaily*. 27 February 1998.

4. Roebke, Joshua. "The Reality Tests." *Seed Magazine*, 4 June 2008. Available at: <http://seedmagazine.com/content/article/the_reality_tests/P3/> [accessed 25 June 2017].

5. By Stannered via Wikimedia Commons. Available at: <https://commons.wikimedia.org/wiki/File:Doubleslit.svg>

6. Available at: <https://en.wikipedia.org/wiki/Delayed_choice_quantum_eraser> [accessed 25 June 2017].

7. Crease, Robert P. "The Most Beautiful Experiment." *Physics World*, 1 September 2002. Available at: <http://physicsworld.com/cws/article/print/2002/sep/01/the-most-beautiful-experiment> [accessed 25 June 2017].

8. Kofler, Johannes, Tomasz Paterek, and Caslav Brukner. "Experimenter's Freedom in Bell's Theorem and Quantum Cryptography." *Physical Review A*, 73(2), November 2005. Available at: <http://arxiv.org/pdf/quant-ph/0510167.pdf> [accessed 25 June 2017].

9. Roebke, J., ref. 4.

10. Hensen, et al. "Loophole-free Bell inequality violation using electron spins separated by 1.3 kilometres." *Nature*. 526: pp. 682–686.

11. Giustina, Marissa; Versteegh, Marijn A. M.; Wengerowsky, Soeren; Handsteiner, Johannes; Hochrainer, Armin; Phelan, Kevin; Steinlechner, Fabian; Kofler, Johannes; Larsson, Jan-Ake; Abellan, Carlos; Amaya, Waldimar; Pruneri, Valerio; Mitchell, Morgan W.; Beyer, Joern; Gerrits, Thomas; Lita, Adriana E.; Shalm, Lynden K.; Nam, Sae Woo; Scheidl, Thomas; Ursin, Rupert; Wittmann, Bernhard; Zeilinger, Anton (2015). "A significant-loophole-free test of Bell's theorem with entangled photons." *Phys. Rev. Lett.* 115, 250401 (2015). Available at: <https://arxiv.org/abs/1511.03190> [accessed 25 June 2017].

12. Shalm, Lynden K.; Meyer-Scott, Evan; Christensen, Bradley G.; Bierhorst, Peter; Wayne, Michael A.; Stevens, Martin J.; Gerrits, Thomas; Glancy, Scott; Hamel, Deny R.; Allman, Michael S.; Coakley, Kevin J.; Dyer, Shellee D.; Hodge, Carson; Lita, Adriana E.; Verma, Varun B.; Lambrocco, Camilla; Tortorici, Edward; Migdall, Alan L.; Zhang, Yanbao; Kumor, Daniel R.; Farr, William H.; Marsili, Francesco; Shaw, Matthew D.; Stern, Jeffrey A.; Abellán, Carlos; Amaya, Waldimar; Pruneri, Valerio; Jennewein, Thomas; Mitchell, Morgan W.; Kwiat, Paul G.; Bienfang, Joshua C.; Mirin, Richard P.; Knill, Emanuel; Nam, Sae Woo (2015). "A strong loophole-free test of local realism." *Phys. Rev. Lett.* 115, 250402 (2015). Available at: <https://www.quantamagazine.org/the-evolutionary-argument-against-reality-20160421/> [accessed 25 June 2017].

13. Gefter, Amanda. "The Evolutionary Argument Against Reality." *Quanta Magazine*, 21 April 2016. Available at: <https://www.quantamagazine.org/20160421-the-evolutionary-argument-against-reality/> [accessed 25 June 2017].

14. Talbot, Michael. *The Holographic Universe* (New York: HarperCollins, 1991) p. 240.

15. Shared Crossing Project. "Shared Death Experience." Available at: <http://www.sharedcrossing.com/shared-dea

th-experience.html> [accessed 25 June 2017].

16. Knapton, Sarah. "First hint of 'life after death' in biggest ever scientific study." *The Telegraph*, 7 October 2014. Available at: <http://www.telegraph.co.uk/science/2016/03/12/first-hint-of-life-after-death-in-biggest-ever-scientific-study/> [accessed 25 June 2017].

17. Beauregard, Mario. *Brain Wars: The Scientific Battle Over the Existence of the Mind and the Proof That Will Change the Way We Live Our Lives* (San Francisco: HarperOne, 2012) pp. 157–160.

18. Van Lommel, Dr. Pim. *Consciousness Beyond Life: The Science of the Near-Death Experience* (San Francisco: HarperOne, 2011) p 111.

19. Van Lommel, pp. 135–158.

20. Schwartz, Gary. *The Afterlife Experiments: Breakthrough Scientific Evidence of Life After Death* (New York: Atria Books, 2003).

21. Alexander, Dr. Eben III. *Proof of Heaven: A Neurosurgeon's Journey into the Afterlife*, 1 edn. (New York: Simon & Schuster, 2012).

22. Alexander, Eben III. "My Experience in Coma," *AANS Neurosurgeon*: Features, Volume 21, Number 2, 2012.

23. Kelly, Edward F., Emily Williams Kelly, Adam Crabtree, Alan Gauld, Michael Grosso, and Bruce Greyson. *Irreducible Mind: Toward a Psychology for the 21st Century* (Lanham, MD: Rowman & Littlefield Publishers, 2007).

24. Kelly et al, p. 421.

25. Hughes, Virginia, "How the Blind Dream," *National Geographic*, 26 February 2014. <http://phenomena.natio nalgeographic.com/2014/02/26/how-the-blind-dream/>

26. Ring, Dr. Kenneth. *Mindsight: Near-Death and Out-of-Body Experiences in the Blind*, edn. 2 (New York: iUniverse, 2008).

27. Shared Crossing Project.

28. Lawrence, Madelaine and Elizabeth Repede. "The Incidence

of Deathbed Communications and Their Impact on the Dying Process." *American Journal of Hospice & Palliative Medicine*, November 2013, vol. 30, no. 7, pp. 632–639.

29. Wills-Brandon, Dr Carla, "The Trigger of Deathbed Visions: Dr. Carla. Wills-Brandon's Research." <http://www.near-death.com/experiences/triggers/deathbed-visions.html>

30. Monroe, Robert. *Ultimate Journey* (New York: Broadway Books, 1994) p. 2.

31. Smith, Andra M. and Claude Messier. "Voluntary out-of-body experience: an fMRI study," *Frontiers in Human Neuroscience*, 10 February 2014. Available at: <http://journal.frontiersin.org/article/10.3389/fnhum.2014.00070/full> [accessed 25 June 2017].

32. Kushins, Jordan. "Scientists unlock mystery of out-of-body experiences." *Gizmodo*, 7 March 2014. Available at: <http://sploid.gizmodo.com/scientists-unlock-mystery-of-womanwho-sees-herself-out-1538196076> [accessed 25 June 2017].

33. Madhudvisa dasa, "The History of Bhagavad Gita," Krishna.org, 6 August 2015 <http://krishna.org/the-history-of-bhagavad-gita/> [accessed 25 June 2017].

34. Page, Michael, *Cheating Death Twice: Confessions of a Soldier* (Cirencester, Gloucestershire: Mereo, 2013) p. 68.

35. Julius Caesar, *De Bello Gallico*, VI.

36. Murray, Margaret Alice, *The Splendor That Was Egypt: A General Survey of Egyptian Culture and Civilization* (Whitefish, MT: Literary Licensing, LLC, 2013).

37. Von Ward, Paul, "The Reincarnation Experiment," 2008. Available at: <http://www.reincarnationexperiment.org/reincarnationhistory.html> [accessed 25 June 2017].

38. Von Ward.

39. Rooke, Andrew, "Reincarnation in African Traditional Religion," *Sunrise Magazine*, November 1980. Available at: <http://www.theosophy-nw.org/theosnw/world/africa/af-

rook2.htm> [accessed 25 June 2017].

40. Hays, Jeffrey, "Shamanism in Russia and Mongolia," 2008. Available at: <http://factsanddetails.com/world/cat55/sub 350/item1919.html> [accessed 25 June 2017].

41. "Traditional Native Concepts of Death," 1 September 2014. Available at: <http://nativeamericannetroots.net/diary/1726> [accessed 25 June 2017].

42. Crisp, Tony, "Australian Aborigine Dream Beliefs." Available at: <http://dreamhawk.com/dream-encyclopedia/australian-aborigine-dream-beliefs/> [accessed 25 June 2017].

43. Meyer, Jacques A., *Embrace the World: A compelling blend of wisdom from history's great minds. Thundering fun ride of inner peace* (New York: iUniverse, Inc.) p. 33.

44. Von Ward.

45. Von Ward.

46. Asgharzadeh, Reza, "Re-incarnation in the Qur'an," 22 January 2012. Available at: <https://reincarnationquran. wordpress.com/2012/01/> [accessed 25 June 2017].

47. Milligan, Wes, "The Reincarnation Story & Past Life Memories of James Leininger," *Acadiana Profile Magazine*, December 2004. Available at: <http://www.iisis.net/index. php?page=james-huston-james-leininger-reincarnaton-wes-milligan-acadiana-profile&hl=en_US> [accessed 25 June 2017].

48. MacIsaac, Tara, "Nazi Airman Reincarnated as English Railway Worker? Stunning Coincidences Suggest So," *Epoch Times*, 22 November 2014. Available at: <http://www. theepochtimes.com/n3/1097777-nazi-pilot-reincarnated-as-english-railway-worker-stunning-coincidences-suggest-so/> [accessed 25 June 2017].

49. Weiss, Brian, *Many Lives, Many Masters* (New York: Fireside, 1988).

50. Targ, Russell and Jane Katra, "The Scientific and Spiritual Implications of Psychic Abilities." Available at: <http://

www.espresearch.com/espgeneral/doc-AT.shtml> [accessed 25 June 2017].

51. Targ, Russell, *The Reality of ESP* (Wheaton, IL: Quest Books, 2012).

52. Targ, R., ref. 51.

53. Jahn, R. G., B. J. Dunne, R. D. Nelson, Y. H. Dobyns, and G. J. Bradish. "Correlations of Random Binary Sequences with Pre-Stated Operator Intention: A Review of a 12-Year Program." *Journal of Scientific Exploration*, Vol. 11, No. 3: pp. 345–367, 1997.

54. Jahn, Robert G. "The Persistent Paradox of Psychic Phenomena: An Engineering Perspective." *IEEE*, Proceedings, 70: pp. 136, 1982.

55. Parker, Adrian. "Report on Work in Progress on the Ganzfeld Project—January1996–June1997." University of Göteborg Department of Psychology. Available at: <http://parapsykologi.se/2012/09/om-ganzfeld-i-goteborg/> [accessed 25 June 2017].

56. Sheldrake, Rupert. "Experiments on the Sense of Being Stared at: The Elimination of Possible Artifacts," *Journal of the Society for Psychical Research*, Vol. 65, 2001: pp. 122–137. Available at: <https://www.sheldrake.org/research/sense-of-being-stared-at/experiments-on-the-sense-of-being-stared-at-eliminating-artefacts> [accessed 25 June 2017].

57. Delanoy, D. "Experimental Evidence Suggestive of Anomalous Consciousness Interactions." 2nd Gauss Symposium, Munich, August 1993. Available at: <http://www.tcm.phy.cam.ac.uk/~bdj10/psi/delanoy/node5.html#SECTION00050000000000000000> [accessed 25 June 2017].

58. Radin, Dean, *The Conscious Universe: The Scientific Truth of Psychic Phenomena* (San Francisco: HarperEdge, 1997) p. 89.

59. Lowery, George, "Study showing that humans have some psychic powers caps Daryl Bem's career," *Cornell Chronicle*, 6 December 2010. Available at: <http://www.news.cornell.

edu/stories/2010/12/study-looks-brains-ability-see-future>
[accessed 25 June 2017].

60. Weiss, Charles, "Expressing Scientific Uncertainty," *Law, Probability and Risk*, (2003) pp. 2, 25–46.

61. Available at: <https://upload.wikimedia.org/wikipedia/commons/0/09/Sibling-correlation-422.png> [accessed 25 June 2017].

62. Hoyle, Fred, *The Intelligent Universe* (New York: Holt, Rinehart and Winston, 1988) p. 17.

63. Paul, Marla, "Your Memory is like the Telephone Game." Northwestern University, 19 September 2012. Available at: <http://www.northwestern.edu/newscenter/stories/2012/09/your-memory-is-like-the-telephone-game.html> [accessed 25 June 2017].

64. Available at: <http://www.bbc.com/news/entertainment-arts-29160096> [accessed 25 June 2017].

65. Talbot, Michael, *The Holographic Universe* (New York: HarperCollins, 1991) p. 31.

66. "No Brain? No Problem! Slime Mold Can Learn," *Seeker*, 26 April 2016.

67. "Slime Mold 'Biocomputer' Maps Ancient Roman Roads," *Seeker*, 11 March 2015.

68. Nelson, Roger D. "Wishing for Good Weather: A Natural Experiment in Group Consciousness," *Journal of Scientific Exploration*, Vol. 11, No. 1, pp. 47–58, 1997.

69. "Global Correlations in Random Data." 23 July 2005. Available at: <http://noosphere.princeton.edu/> [accessed 25 June 2017].

70. MacIsaac, Tara, "Evidence Group Consciousness May Have a Physical Effect on the World," *Epoch Times*, 16 June 2015.

71. Thalbourne, Michael A., "Science Versus Showmanship: A History of the Randi Hoax," *The Journal of the American Society for Psychical Research*, Vol. 89, October 1995.

Chapter 7

1. "We are not Alone: Scientists conclude whales, dolphins and many other species are conscious," *Whales and Dolphins Conservation*, 14 August 2012. Available at: <http://us.whales.org/blog/2012/08/we-are-not-alone-scientists-conclude-whales-dolphins-and-many-other-species-are/> [accessed 25 June 2017].

2. Butler, A. B. "Hallmarks of Consciousness." *Advances in Experimental Medicine and Biology*, 2012; 739: pp. 291–309. doi: 10.1007/978-1-4614-1704-0_19

3. Laszlo, Dr. Ervin, "16 Hallmarks of Consciousness," GlobalWoman.com, 1 October 2015. Available at: <http://www.globalwoman.co/2015/10/16-hallmarks-of-oneness-consciousness-dr-ervin-laszlo/> [accessed 25 June 2017].

4. Eells, Josh, "He talked to plants. And they talked back," *The New York Times Magazine*, 21 December 2013. Available at: <http://www.nytimes.com/news/the-lives-they-lived/2013/12/21/cleve-backster/> [accessed 25 June 2017].

5. "How plants warn each other of danger," BBC News, 7 February 2012. Available at: <http://www.bbc.co.uk/news/science-environment-16916474> [accessed 25 June 2017].

6. Alleyne, Richard, "Women's voices 'make plants grow faster' finds Royal Horticultural Society," *The Telegraph*, 22 June 2009. Available at: <http://www.telegraph.co.uk/news/earth/earthnews/5602419/Womens-voices-make-plants-grow-faster-finds-Royal-Horticultural-Society.html> [accessed 25 June 2017].

7. Pollan, Michael, "The Intelligent Plant," *The New Yorker*, 23/30 December 2013. Available at: <http://www.newyorker.com/magazine/2013/12/23/the-intelligent-plant> [accessed 25 June 2017].

8. MacIsaac, Tara, "Do Inanimate Objects Have Thoughts and Feelings?" *Epoch Times*, 5 August 2014. Available at: <http://www.theepochtimes.com/n3/845646-do-inanimate-objects-

have-thoughts-and-feelings/> [accessed 25 June 2017].

9. Koch, Christof, "Is Consciousness Universal?" *Scientific American*, 1 January 2014. Available at: <https://www. scientificamerican.com/article/is-consciousness-universal/> [accessed 25 June 2017].

10. Koch, Christof, ref. 9.

11. Available at: <http://www.chinabuddhismencyclopedia. com/en/index.php/Sentient_beings> [accessed 25 June 2017].

12. O'Flaherty, Wendy Doniger. *Dreams, Illusion, and Other Realities* (Chicago, Il: University of Chicago Press, 1986) p. 119.

13. Robsville, Sean, "Consciousness and mind are not emergent phenomena or emergent properties of the brain," *Transcultural Buddhism*, 29 October 2009. Available at: <http://seanrobsville.blogspot.com/2009/10/consciousness-and-mind-as-emergent.html> [accessed 25 June 2017].

14. Robsville, Sean, ref. 13.

15. Chalmers, David J., "Thoughts on Emergence," 6 October 1990. Available at: <http://consc.net/notes/emergence.html> [accessed 25 June 2017].

16. Merritt, Dr. John L. and J. Lawrence Merritt, II. *When Does Human Life Begin* (Kenmore, WA: Crystal Clear Books, 2012).

17. Swami Kriyananda (J. Donald Walters), *Conversations with Yogananda* (Nevada City, CA: Crystal Clarity Publishers, 2004) p. 206.

18. Aristotle. *History of Animals*, Book VII, Chapter 3, 583b.

19. Khitamy, Badawy A. B., "Divergent Views on Abortion and the Period of Ensoulment," *Sultan Qaboos University Medical Journal*, February 2013; 13(1): pp. 26–31.

20. Hasgul, Olgun, Ensoulment: "When Does Human Life Begin?" *The Fountain*, Issue 50: April–June 2005.

21. Monroe, Robert, *Far Journeys* (New York: Broadway Books, 2001) pp. 140–141.

Chapter 8

1. Feynman, Richard, at his Nobel Lecture for his Nobel Prize in Physics on 11 December 1965.

2. Teuscher, Christof, *Alan Turing: Life and Legacy of a Great Thinker* (New York: Springer, 2006) p. 54.

3. Fischer, M. C., B. Gutierrez-Medina, and M. G. Raizen, "Observation of the Quantum Zeno and Anti-Zeno effects in an unstable system," Department of Physics, the University of Texas at Austin, Texas, 1 February 2008.

4. Inglis-Arkell, Esther, "An experiment that might let us control events millions of years ago," *io9*, 19 February 2014. Available at: <http://io9.gizmodo.com/an-experiment-that-might-let-us-control-events-millions-1525760859> [accessed 25 June 2017].

5. Available at: <https://commons.wikimedia.org/wiki/File:Kim_EtAl_Quantum_Eraser.svg> [accessed 25 June 2017].

6. Kim, Yoon-Ho; R. Yu; S. P. Kulik; Y. H. Shih; Marlan Scully (2000). "A Delayed 'Choice' Quantum Eraser." *Physical Review Letters* 84: pp. 1–5. 2000. Available at: <https://arxiv.org/abs/quant-ph/9903047> [accessed 25 June 2017].

7. "News from the world of quantum physics: A non-causal quantum eraser," University of Vienna, 9 January 2013. Available at: <https://www.sciencedaily.com/releases/2013/01/130109105932.htm> [accessed 25 June 2017].

8. "Experiment confirms quantum theory weirdness," Phys.org, 27 May 2015. Available at: <http://phys.org/news/2015-05-quantum-theory-weirdness.html> [accessed 25 June 2017].

9. Davies, Paul. "Laying down the laws," *New Scientist*, 30 June–6 July 2007: pp. 30–34.

10. "Scientific Benefits of Meditation," *Mental Health Daily*. Available at: <http://mentalhealthdaily.com/2015/03/26/scientific-benefits-of-meditation-list/> [accessed 25 June 2017].

11. Saffer, Dan, "Review: The User Illusion," KickerStudio, 12 February 2009. Available at: <http://www.kickerstudio.com/2009/02/review-the-user-illusion/> [accessed 25 June 2017].

12. Porter, Jane, "You're More Biased Thank You Think," *Fast Company*, 6 October 2014.

13. Moscowitz, Clara, "Mind's Limit Found: 4 Things at Once," *Live Science*, 27 April 2008. Available at: <http://www.livescience.com/2493-mind-limit-4.html> [accessed 25 June 2017].

14. "Lotus Guide: An Interview with Bruce Lipton," 18 September 2012. Available at: <http://www.brucelipton.com/resource/interview/lotus-guide-interview-bruce-lipton> [accessed 25 June 2017].

15. Dispenza, Joe, from *What the Bleep Do We Know*, Dir. William Arntz. (New York: Samuel Goldwyn Films, 2004). Film.

16. Reber, Paul, "What is the Memory Capacity of the Human Brain," *Scientific American*, 1 May 2010. Available at: <http://www.scientificamerican.com/article.cfm?id=what-is-the-memory-capacity> [accessed 25 June 2017].

17. Srinivasan, Thaiyar M., "Genetics, epigenetics, and pregenetics," *International Journal of Yoga*. July–December 2011; 4(2): pp. 47–48.

18. Hart, Eva, *A Girl Aboard the Titanic* (Stroud, UK: Amberley, 2012).

19. "Did Abraham Lincoln predict his own death?" History.com, 31 October 2012. Available at: <http://www.history.com/news/ask-history/did-abraham-lincoln-predict-his-own-death> [accessed 25 June 2017].

20. Purdy, Mike, "Both Lincoln and Kennedy Had Premonitions of Their Assassinations," PresidentialHistory.com, 28 January 2013. Available at: <http://presidentialhistory.com/2013/01/both-lincoln-and-kennedy-had-premonitions-of-their-assassinations.html> [accessed 25 June 2017].

21. Available at: <http://abcnews.go.com/US/story?id=91891&pa ge=1> [accessed 25 June 2017].

22. Bem, Daryl J., "Feeling the Future: Experimental Evidence for Anomalous Retroactive Influences on Cognition and Affect," *Journal of Personality and Social Psychology*, 2011, pp. 100, 407–425.

23. Aldhous, Peter, "Is this evidence that we can see the future?" *New Scientist*, 11 November 2010. Available at: <https:// www.newscientist.com/article/dn19712-is-this-evidence- that-we-can-see-the-future> [accessed 25 June 2017].

24. Aldhous, Peter, ref. 23.

25. Available at: <http://deanradin.blogspot.com/2010/09/ feeling-future.html> [accessed 25 June 2017].

26. Lipka, Michael, "18% of Americans say they've seen a ghost," Pew Research Center, 30 October 2015. Available at: <http://www.pewresearch.org/fact-tank/2015/10/30/18- of-americans-say-theyve-seen-a-ghost/> [accessed 25 June 2017].

27. Harish, Alon, "UFOs Exist, Say 36 Percent in National Geographic Survey," ABC News, 27 June 2012. Available at: <http://abcnews.go.com/Technology/ufos-exist-americans- national-geographic-survey/story?id=16661311> [accessed 25 June 2017].

28. "Poll: U.S. hiding knowledge of aliens," CNN Interactive, 15 June 1997. Available at: <http://www.cnn.com/US/9706/15/ ufo.poll/> [accessed 25 June 2017].

29. Jacobs, Ryan, "Why So Many Icelanders Still Believe in Invisible Elves," *The Atlantic*, 29 October 2013. Available at: <http://www.theatlantic.com/international/archive/2013/10/ why-so-many-icelanders-still-believe-in-invisible- elves/280783/> [accessed 25 June 2017].

30. Available at: <http://www.unknownexplorers.com/lochnes smonster.php> [accessed 25 June 2017].

31. Available at: <http://www.ufoevidence.org/documents/

doc1744.htm> [accessed 25 June 2017].

32. Lindsay, E. M., "Maskelyne and Meteors," *Irish Astronomical Journal*, vol. 8(3), p. 69.

33. Smith, Jonathan C., *Pseudoscience and Extraordinary Claims of the Paranormal* (Malden, MA: Wiley-Blackwell, 2010).

34. MacDonald, Fiona, "Humans didn't even see the colour blue until modern times, research suggests," ScienceAlert, 4 March 2015. Available at: <http://www.sciencealert.com/humans-couldn-t-even-see-the-colour-blue-until-modern-times-research-suggests> [accessed 25 June 2017].

35. "Srimad Bhagavatam: Canto 10—Chapter 87—Text 41." *Srimad Bhagavatam (Bhagavata Purana); the Story of Krishna.* N.p., n.d. Web. 17 November 2016.

36. Plaga, R. (1997). "On a possibility to find experimental evidence for the many-worlds interpretation of quantum mechanics." *Foundations of Physics.* 27: pp. 559–577.

37. Rae, Alastair I. M. and Sidney L. Harring, *Quantum Physics: Illusion or Reality?* (Cambridge, UK: Cambridge University Press, 1994) p. 79.

38. Arkani-Hamed, N., Dimopoulos, S. and Dvali, G. "The Universe's Unseen Dimensions," *Scientific American*, vol. 12, no. 2. 2002: pp. 66–73.

39. Lennon, Jen, "Mysterious Time Travelers with Convincing Stories," Ranker. Available at: <http://www.ranker.com/list/time-traveler-stories/jenniferlennon> [accessed 25 June 2017].

40. Schwarz, Rob, "Sir Victor Goddard's Time Slip Adventure," *Stranger Dimensions*, 25 February 2015. Available at: <http://www.strangerdimensions.com/2015/02/25/sir-victor-goddards-time-slip-adventure/> [accessed 25 June 2017].

41. Stevenson, Ian, "A Preliminary Report of a New Case of Responsive Xenoglossy: The Case of Gretchen," *The Journal of the American Society for Psychical Research*, Vol. 70, January 1976.

Chapter 10

1. Roser, Max, "Homicides," *Our World in Data*. Available at: <https://ourworldindata.org/homicides/> [accessed 25 June 2017].

About the Author

Jim Elvidge holds a Master's Degree in Electrical Engineering from Cornell University, New York. He has applied his training in the high-tech world as a leader in technology and enterprise management, including many years in executive roles for various companies and entrepreneurial ventures. He also holds four patents in digital signal processing.

Beyond the high-tech realm, however, Elvidge has years of experience as a musician, writer and truth-seeker. He merged his technology skills with his love of music, developed one of the first PC-based digital music samplers and cofounded RadioAMP, the first private-label online-streaming radio company.

For many years, Elvidge has kept pace with the latest research, theories and discoveries in the varied fields of subatomic physics, cosmology, artificial intelligence, nanotechnology and the paranormal. This unique knowledge base provided the foundation for his first full-length book, *The Universe — Solved!* Additional years of research, inward journeys, and dialog with other truth-seekers and reality researchers have given him the foundation for this book, *Digital Consciousness*.

BOOKS

Iff Books

ACADEMIC AND SPECIALIST

Iff Books publishes non-fiction. It aims to work with authors and titles that augment our understanding of the human condition, society and civilisation, and the world or universe in which we live.

If you have enjoyed this book, why not tell other readers by posting a review on your preferred book site.

Recent bestsellers from Iff Books are:

Why Materialism Is Baloney
How True Skeptics Know There is no Death and Fathom Answers
to Life, the Universe, and Everything
Bernardo Kastrup
A hard-nosed, logical, and skeptic non-materialist metaphysics,
according to which the body is in mind, not mind in the body.
Paperback: 978-1-78279-362-5 ebook: 978-1-78279-361-8

The Fall
Steve Taylor
The Fall discusses human achievement versus the issues of war,
patriarchy and social inequality.
Paperback: 978-1-90504-720-8 ebook: 978-184694-633-2

Brief Peeks Beyond
Critical Essays on Metaphysics, Neuroscience, Free Will,
Skepticism and Culture
Bernardo Kastrup
An incisive, original, compelling alternative to current mainstream
cultural views and assumptions.
Paperback: 978-1-78535-018-4 ebook: 978-1-78535-019-1

Framespotting
Changing How You Look at Things Changes How
You See Them
Laurence & Alison Matthews
A punchy, upbeat guide to framespotting. Spot deceptions and
hidden assumptions; swap growth for growing up. See and be free.
Paperback: 978-1-78279-689-3 ebook: 978-1-78279-822-4

Is There an Afterlife?

David Fontana

Is there an Afterlife? If so what is it like? How do Western ideas of the afterlife compare with Eastern? David Fontana presents the historical and contemporary evidence for survival of physical death.

Paperback: 978-1-90381-690-5

Nothing Matters

A Book About Nothing

Ronald Green

Thinking about Nothing opens the world to everything by illuminating new angles to old problems and stimulating new ways of thinking.

Paperback: 978-1-84694-707-0 ebook: 978-1-78099-016-3

Panpsychism

The Philosophy of the Sensuous Cosmos

Peter Ells

Are free will and mind chimeras? This book, anti-materialistic but respecting science, answers: No! Mind is foundational to all existence.

Paperback: 978-1-84694-505-2 ebook: 978-1-78099-018-7

Punk Science

Inside the Mind of God

Manjir Samanta-Laughton

Many have experienced unexplainable phenomena; God, psychic abilities, extraordinary healing and angelic encounters. Can cutting-edge science actually explain phenomena previously thought of as 'paranormal'?

Paperback: 978-1-90504-793-2

The Vagabond Spirit of Poetry

Edward Clarke

Spend time with the wisest poets of the modern age and of the past, and let Edward Clarke remind you of the importance of poetry in our industrialized world.

Paperback: 978-1-78279-370-0 ebook: 978-1-78279-369-4

Readers of ebooks can buy or view any of these bestsellers by clicking on the live link in the title. Most titles are published in paperback and as an ebook. Paperbacks are available in traditional bookshops. Both print and ebook formats are available online.

Find more titles and sign up to our readers' newsletter at http://www.johnhuntpublishing.com/non-fiction

Follow us on Facebook at https://www.facebook.com/JHPNonFiction and Twitter at https://twitter.com/JHPNonFiction